Sue B. Moon   Renata Teixeira
Steve Uhlig (Eds.)

Passive and Active Network Measurement

10th International Conference, PAM 2009
Seoul, Korea, April 1-3, 2009
Proceedings

 Springer

Volume Editors

Sue B. Moon
KAIST, Computer Science Department
Daejeon 305-701, Republic of Korea
E-mail: sbmoon@kaist.edu

Renata Teixeira
CNRS - Centre National de la Recherche Scientifique
and
Université Pierre et Marie Curie
Paris 6, 75016 Paris, France
E-mail: renata.teixeira@lip6.fr

Steve Uhlig
T-labs/TU Berlin
Berlin, Germany
E-mail: steve@net.t-labs.tu-berlin.de

Library of Congress Control Number: Applied for

CR Subject Classification (1998): C.2, C.4, H.4, K.6.5

LNCS Sublibrary: SL 5 – Computer Communication Networks and Telecommunications

ISSN	0302-9743
ISBN-10	3-642-00974-3 Springer Berlin Heidelberg New York
ISBN-13	978-3-642-00974-7 Springer Berlin Heidelberg New York

springer.com

© Springer-Verlag Berlin Heidelberg 2009
Printed in Germany

Typesetting: Camera-ready by author, data conversion by Scientific Publishing Services, Chennai, India
Printed on acid-free paper   SPIN: 12636129   06/3180   5 4 3 2 1 0

Lecture Notes in Computer Science 5448

Commenced Publication in 1973
Founding and Former Series Editors:
Gerhard Goos, Juris Hartmanis, and Jan van Leeuwen

Editorial Board

Preface

The 2009 edition of the Passive and Active Measurement Conference was the tenth of a series of successful events. Since 2000, the Passive and Active Measurement (PAM) conference has provided a forum for presenting and discussing innovative and early work in the area of Internet measurement. This event focuses on research and practical applications of network measurement and analysis techniques. The conference's goal is to provide a forum for current work in its early stages. This year's conference was held at Seoul National University in Seoul, the 600-year-old capital of Korea.

PAM 2009 attracted 77 submissions. Each paper was carefully reviewed by at least three members of the Technical Program Committee. The reviewing process led to the acceptance of 22 papers and 2 demos. Demos are a novelty of this year's PAM. The goal of demos is to present measurement tools, which can be so useful for our community. The papers and demos were arranged into nine sessions covering the following areas: routing and forwarding; topology and delay; methods for large-scale measurements; wireless; management tools; audio and video traffic; peer-to-peer; traffic measurements; and measurements of anomalous and unwanted traffic. The technical program of the conference was complemented by a half-day PhD student workshop with poster presentations and a panel.

We would like to thank all members of the Technical Program Committee for their timely and thorough reviews. Special thanks to Balachander Krishnamurthy and Konstantina Papagiannaki for handling all papers with PC-Chair conflict. We would also like to thank Sojin Lee for laying out plans for the budget, lodging, and banquets and seeing them through, as well as Seoyeon Kang, who managed the website and was always there to help out with last-minute details.

Last but not least, we are extremely grateful to Korea Research Foundation, Intel, Endace, and Telefonica, whose sponsoring allowed us to keep registration costs low and to offer several travel grants to PhD students.

April 2009

Sue Moon
Renata Teixeira

Organization

Organization Committee

General Chair	Sue Moon (KAIST, South Korea)
Program Chair	Renata Teixeira (CNRS and UPMC Paris Universitas, France)
Publication Chair	Steve Uhlig (TU Berlin/T-labs, Germany)
Local Arrangements Chair	Taekyoung Kwon (Seoul National University, South Korea)
Finance Chair	Sojin Lee (KAIST, South Korea)

Steering Committee

Nevil Brownlee	University of Auckland
Mark Claypool	Worcester Polytechnic Institute
Ian Graham	Endace
Sue Moon	KAIST
Konstantina Papagiannaki	Intel Research Pittsburgh
Renata Teixeira	CNRS and UPMC Paris Universitas
Michael Rabinovich	Case Western Reserve University
Steve Uhlig	TU Berlin/T-labs

Program Committee

Jussara Almeida	Universidade Federal de Minas Gerais, Brazil
Ernst Biersack	Eurecom, France
Kenjiro Cho	WIDE/IIJ, Japan
kc claffy	CAIDA, USA
Mark Crovella	Boston University, USA
Anja Feldmann	TU Berlin/T-labs, Germany
Clarence Filsfils	Cisco, Belgium
Jaeyeon Jung	Intel Research Seattle, USA
Thomas Karagiannis	Microsoft Research Cambridge, UK
Balachander Krishnamurthy	AT&T, USA
Anukool Lakhina	Guavus, India
Simon Leinen	Switch, Switzerland
Olaf Maennel	TU Berlin/T-labs, Germany
Z. Morley Mao	University of Michigan, USA
Priya Mahadevan	HP Labs, USA
Maurizio Molina	Dante, UK
Hung Nugyen	University of Adelaide, Australia

Konstantina Papagiannaki	Intel Pittsburgh, USA
Vern Paxson	UC Berkeley, USA
Himabindu Pucha	Carnegie Mellon University, USA
Jennifer Rexford	Princeton University, USA
Renata Teixeira	CNRS and UPMC Paris Universitas, France
Jia Wang	AT&T Labs Research, USA
Tanja Zseby	Fraunhofer Institute Fokus, Germany

External Reviewers

Cristian Estan	University of Wisconsin
Nick Feamster	Georgia Tech
Kirill Levchenko	UCSD
Matthew Roughan	University of Adelaide
Moritz Steiner	Institut Eurecom
Geoffrey M.Voelker	UCSD

Sponsoring Institutions

Korea Research Foundation
Intel
Endace
Telefonica

Table of Contents

Characterization of Routing and Forwarding

Topology and Delay

Methods for Large-Scale Measurements

Wireless

Management Tools

Audio and Video Traffic

Peer-to-Peer

Traffic Measurements

Measurements of Anomalous and Unwanted Traffic

Characterization of
Routing and Forwarding

Revisiting Route Caching:
The World Should Be Flat

Changhoon Kim[1], Matthew Caesar[2], Alexandre Gerber[3], and Jennifer Rexford[1]

[1] Princeton University,
[2] UIUC,
[3] AT&T Labs–Research

Abstract. Internet routers' forwarding tables (FIBs), which must be stored in expensive fast memory for high-speed packet forwarding, are growing quickly in size due to increased multihoming, finer-grained traffic engineering, and deployment of IPv6 and VPNs. To address this problem, several Internet architectures have been proposed to reduce FIB size by returning to the earlier approach of *route caching*: storing only the working set of popular routes in the FIB. This paper revisits route caching. We build upon previous work by studying flat, uni-class (/24) prefix caching, with modern traffic traces from more than 60 routers in a tier-1 ISP. We first characterize routers' working sets and then evaluate route-caching performance under different cache replacement strategies and cache sizes. Surprisingly, despite the large number of deaggregated /24 subnets, caching uni-class prefixes can effectively curb the increase of FIB sizes. Moreover, uni-class prefixes substantially simplify a cache design by eliminating longest-prefix matching, enabling FIB design with slower memory technologies. Finally, by comparing our results with previous work, we show that the distribution of traffic across prefixes is becoming increasingly skewed, making route caching more appealing.

1 Introduction

Packet forwarding on core Internet routers is an extremely challenging process. Upon receiving an IP packet, routers have just a few nanoseconds to buffer the packet, select the *longest-matching prefix* covering the packet's destination, and forward the packet to the corresponding outbound interface. To allow this process to happen quickly, routers often make use of special-purpose high-speed memory such as TCAM and SRAM. Unfortunately, the need for multi-homing and fine-grained traffic engineering, the desire to mitigate prefix hijacking by advertising more-specific routes, and the continuing rapid growth of the Internet have rapidly increased the number of routes that an Internet router has to maintain. The accelerating deployments of protocols with large address spaces such as IPv6 and VPNs, combined with the rapidly increasing link speeds, stand to worsen this problem even further.

Unfortunately, the special-purpose memory used for high-speed packet forwarding is orders of magnitude more expensive, more power-hungry, bigger in physical size, and generates substantially more heat than conventional DRAM. This is because manufacturing fast memory requires more transistors per bit [1]. Due to these factors, it

S.B. Moon et al. (Eds.): PAM 2009, LNCS 5448, pp. 3–12, 2009.

will be increasingly challenging to manufacture a line card with large, fast memory at a reasonable price and power-consumption budget. To save expenses, service providers may therefore be forced to provision with little headroom for growth, making their networks unable to handle sudden spikes in table size. Providers will also be forced to upgrade their equipment more often, a substantial problem given maintenance of operational networks can be much more expensive than hardware costs. To keep up with these demands, future routers must reduce the number of routes stored in the FIB.

In this paper we revisit *route caching*, where the FIB stores only the frequently-used routes and other routes are retrieved from a larger but slower memory (e.g., DRAM) on a miss. This approach is motivated by the fact that Internet traffic exhibits high degrees of temporal locality (as packets are grouped into flows, which are often transmitted in bursts) and spatial locality (as many hosts access a small number of popular destinations). In fact, route caching *was* once widely used in Internet routers. In the late 1980s and early 1990s, most routers were built with a route caching capability, such as *fast switching* [2, 3]. Unfortunately, these designs were not able to keep up with fast-increasing packet forwarding rates, due to the large cost of cache misses, such as lower throughput and high packet loss ratio. While this limitation of route caching is yet to be addressed, revisiting it with modern Internet traffic seems worthwhile because recent research results indicate that route caching might be both *possible* and *necessary*:

Route caching may be possible: New Internet routing architectures (such as ViAg-gre [4], or SEATTLE [5]) can improve feasibility of route caching by reducing the cost of a cache miss. For example, when a cache miss happens, a router can immediately forward a packet via a backup default route, without forcing the packet to be queued (while waiting for the cache to be updated from the slow memory) or to traverse the "slow path" (through the router CPU). The backup default route indirectly leads the packet to an alternate router that always maintains a correct route to the destination in its FIB (more explanation in Section 2.1). This "fall-back" mechanism has substantial performance benefits, because packets can always be sent immediately after the cache lookup completes. Since a route cache can be much smaller than a full FIB and takes substantially less time for a lookup, ensuring line-rate packet forwarding becomes easier.

Route caching may be necessary: New protocols with larger (e.g., IPv6) or even flat address spaces (e.g., ROFL [6], LISP [7], AIP [8]) have been proposed to facilitate the Internet's growth and configuration. However, deploying these protocols would significantly increase FIB sizes beyond the capacities of much currently-deployed equipment, and is predicted to require several million FIB entries within several years if current growth trends continue. When FIBs fill up, conventional routers crash or begin behaving incorrectly [9], forcing operators to deploy new line cards or even routers with a larger memory. Alternatively, in such a case, the use of caching would only increase the volume of the traffic handled via the "fall-back" mechanism, instead of incurring hard crashes, improving availability and extending times between router upgrades.

We start by describing our traffic traces collected from over 60 routers in a tier-1 ISP, and justify caching flat *uni-class* (i.e., /24) prefixes (Section 2). We then characterize the working sets of popular prefixes (Section 3) and evaluate route-caching performance under our uni-class model (Section 4). Despite the significantly larger number of uni-class prefixes as compared to CIDR, the cache size needed for a reasonably small miss rate is comparable to the size of FIBs in conventional Internet routers. Moreover, a uniform prefix length allows use of *hashing* for faster lookups, greatly reducing the number of memory accesses per lookup (e.g., looking up an item in a chained hash table with N bins and N items requires 1.58 memory accesses on average, whereas a lookup in a trie-based approach, such as [10], takes much more than that). Therefore, caching uni-class prefixes may be implementable with a slower and cheaper memory (e.g., RL-DRAM [1], or even regular DRAM). Finally, by comparing our results with previous work, we show that Internet traffic today is more amenable to caching (Section 5).

2   Measurement Methodology and Route-Cache Design

Our data sets were collected from a tier-1 ISP's backbone in the U.S. First, we collected *unsampled packet-level* traces over a one-week period at an access router servicing regional DSL subscribers in the ISP. Second, we collected *flow-level* traffic records from more than 60 edge routers in different geographical and topological regions. Since the volume of the traffic transited by those routers was extremely large, we utilized a *sampling* technique (Sampled NetFlow [11]) to reduce the overhead of collecting traffic. These two sets of traces are complementary to each other; the flow-level traces allow us to compare behavior across different routers, while the packet-level traces allow us to study finer-grained behavior at a single access router. Additionally, using unsampled packet-level traces allows us to validate the accuracy of our methodology for using sampled flow-level traces to study route caching.

DSL traces: We collected IP packet headers that originated from roughly $20,000$ regional DSL subscribers in the USA using our network traffic monitoring platform. Our monitor is attached to the access router aggregating traffic from subscribers, and was configured to monitor *inbound* traffic sent from DSL subscribers to the rest of the Internet. Note that we deliberately captured inbound traffic because destination addresses accessed by the inbound traffic are much more diverse than those by outbound traffic and are thus very challenging for our route-caching study. We ran our monitor for 8 consecutive days from Feb 29 through Mar 7, 2008, and captured roughly 40 billion packets, corresponding to an average of $\sim 65,000$ packets per second.

NetFlow traces: To study differences in workload across routers, we collected NetFlow records from inbound traffic to the ISP via two representative POPs (Points of Presence) containing over 60 routers, respectively located on the east and west coasts of the USA. To avoid overloading the router CPU, we were forced to configure NetFlow to perform deterministic sampling with a sampling ratio of $1/500$ [11]. The active and inactive time-out values were set to 60 and 15 seconds respectively. We ran NetFlow for 15 hours on January 23 - 24, 2008, collecting the information of ~ 330 billion packets (roughly 100K pkts/sec per edge router on average). Some of our analysis (e.g., estimating cache miss rate) requires packet-level arrival information. To construct packet records from flow records, we post-processed the trace to distribute all counted packets in a flow evenly over the measured duration of the flow. To check for sampling inaccuracies, we generated sampled DSL traces by applying the NetFlow sampling algorithm to our unsampled DSL traces. There was no statistically significant difference between the results acquired with sampled and unsampled traces.

2.1 Route Caching Model

To evaluate the performance of route caching, we define a simple and generic caching architecture. In this architecture, a packet forwarding unit (e.g., a line card, or a forwarding engine) incorporates hierarchical, two-level memory. The first level is a *route cache*, which is embodied by a small, but very fast memory containing only a subset of the entire routes. The second level is a *full routing table*, which is a slow, but large memory storing all routes. Once a packet arrives, the forwarding unit first looks up the packet's destination address in the route cache. If it finds a match, the packet is immediately forwarded based on the lookup result. If not, the forwarding unit forwards the packet using a backup route. The backup route indicates an alternate router, which can be either statically configured or dynamically chosen from a small set of alternate routers via a very simple computation (e.g., hashing) on the packet header. Note that this computation can be done in parallel with the cache lookup without increasing packet forwarding latency. The alternate router finally forwards the packet to the destination via a direct route residing in its FIB; to ensure this, administrators can run a well-provisioned router in each POP that always keeps the entire set of routes in its FIB or employ a routing protocol, such as ViAggre [4] or SEATTLE [5], where each router maintains a small amount of additional routing information in its route cache. Apart from this packet forwarding procedure, the forwarding unit separately updates its route cache by looking up the full routing table. If needed, an existing entry in the cache is evicted based on a cache replacement strategy.

Conventional routers store information about paths to CIDR (variable-length) prefixes in their routing and forwarding tables. Hence, at first glance, it appears that the cache should also store information in the same structure. However, caching CIDR prefixes presents a serious technical challenge arising from the need to perform longest-prefix matching: if multiple CIDR prefixes contain a destination address, the most-specific one – longest-matching prefix (LMP) – must be chosen to forward the packet. Unfortunately, in a route-caching system, only the contents of the cache are referred to when making a packet forwarding decision, and thus the *cache-hiding* problem arises: consider an empty cache, and suppose the full routing table contains two prefixes: 10.1.0.0/16 associated with output interface O1, and 10.1.5.0/24 associated with O2. Suppose a new packet destined to 10.1.2.3 arrives. The router will find the LMP for this destination, which is 10.1.0.0/16, and will install the route [10.1.0.0/16 → O1] into the cache. Now, suppose the next packet the router receives is destined to 10.1.5.6. Then, the router will discover [10.1.0.0/16 →O1] in the cache, and send the packet to O1. This, however, is incorrect because the LMP for 10.1.5.6 is 10.1.5.0/24, and hence the packet must have been sent to O2. Unfortunately, proposed solutions to this problem involve either a complicated data structure with on-the-fly computation to eliminate inter-dependency among prefixes [12], or grouping all prefixes containing a destination address as an atomic unit of caching operation (insertion, deletion, and update). The latter approach leads to cache thrashing because the size of an atomic prefix group in today's FIB can be larger than 25, 000. Worse yet, a cache storing CIDR prefixes still has to perform longest-prefix matching on every lookup.

To avoid these difficulties, we explore an alternative model which we refer to as "flat", *uni-class* caching. This model is identical to the technique that Feldmeier used in

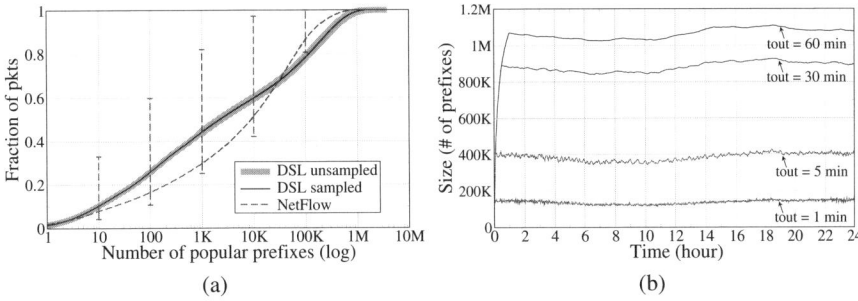

Fig. 1. (a) CDF of uni-class prefix popularity (b) Time series of working set size (DSL trace)

his 1987 study on route caching [13]. In this model, a route cache automatically divides up a CIDR prefix into small, fixed-length (i.e., $/24$) sub-prefixes that are mutually non-overlapping, and then store only the single $/24$ sub-prefix matching to the destination[1]. For example, suppose a full routing table contains a prefix 10.1.0.0/16, along with several more-specific subprefixes under it. In the uni-class model, 10.1.0.0/16 is internally considered as a collection of 256 independent $/24$ routes, ranging from 10.1.0.0/24 to 10.1.255.0/24. Hence, if the destination of an incoming packet is 10.1.2.3, only 10.1.2.0/24 is stored in the route cache. Note that the full routing table still contains a single CIDR prefix, 10.1.0.0/16, as the sub-prefix is generated upon cache update. The reason why we chose $/24$ as the length for our study is multifold. First, it is the most-specific prefix length in common use for inter-domain routing; most providers filter routes more specific than $/24$ to protect themselves against misconfiguration and prefix hijacking. Also, it is the largest source of FIB growth in today's Internet: from Route-Views traces collected from Nov 2001 through Jul 2008, we found that the number of $/24$ prefixes has increased from $60K$ to $140K$, while other prefix lengths increased at a much slower rate (e.g. $/16$s increased from $7K$ to $10K$, $/8$s increased from 18 to 19).

3   Analysis of Traffic Workload

Prefix popularity: Before directly evaluating performance of caching on our traces, we first analyze relevant properties of the workload. Specifically, since our uni-class model can significantly inflate the number of unique prefixes, we are interested in figuring out how small (or large) the set of frequently-accessed $/24$ prefixes is. Figure 1a plots the fraction of packets sent to a given top-x number of prefixes, for both DSL and NetFlow traces. We set the maximum value of the x-axis to be the maximum possible FIB size (9.3M), which is the number of unique $/24$ prefixes we count when deaggregating the 305K CIDR prefixes advertised in the Internet as of February 2008. We find that roughly one tenth (i.e., 0.93M) of the entire prefixes accounts for more than 97% of traffic (consistent with previous work [13, 14, 15]), and nearly 60% (i.e., 5.3M/9M) of the prefixes are never accessed. We found that this result holds across a variety of routers in

---

[1] Uni-class caching can still support more specific routes than $/24$, if desired, by maintaining a small secondary table for those routes.

the network, as shown by the *NetFlow* curve and error bars. Finally, we also confirmed that packet sampling does not noticeably affect the popularity distribution, shown by the *DSL sampled* curve closely matching the *DSL unsampled* curve.

Temporal analysis of working sets: Understanding the temporal dynamics of traffic destinations is important because it determines the stability and predictability of caching performance. Hence, we study how prefix popularity varies at different times and across different timescales as well. To address this, we leverage the notion of a *working set*, which is defined to be the set of prefixes accessed over a given period of time. The size of the working set and its variation are often used to estimate how large a cache will be needed to achieve a low miss rate. Figure 1b shows variation in the size of the working set over time, under four different definitions of the working set: the set of items accessed over last 60, 30, 5, and 1 minute. As one might expect, larger *tout* values induce larger working set sizes. Interestingly, however, we found the size of working sets across a variety of timeout values to be highly stable. Over our entire set of traces, the standard deviation in working set size was $\sim 3.2\%$ of the average, and the maximum working set size was no more than $\sim 5.6\%$ larger than the average. This fact bodes well for deployability of caching solutions using small fixed-size caches, as cache sizes can be provisioned close to the observed mean without requiring large headroom. Our further analysis also confirmed that the contents of the working sets vary little.

Cross-router analysis of working sets: Understanding the similarity of working sets across routers is important because it allows a cache to be *pre-provisioned* with the contents from another cache, significantly reducing the cold misses after network events (e.g., router or line-card reboot, routing change due to a network failure). Thus, we need to understand how working sets vary across different routers, router roles, and regions within the ISP. We chose several representative access routers: 7 from the ISP's west coast POP, and 9 from its east coast POP. To quantify the similarity of working sets at two different routers r_1 and r_2, we define the *similarity factor* (sfactor) as the number of common prefixes maintained by both r_1 and r_2, divided by the sum of the number prefixes maintained individually by r_1 and r_2. We computed the *sfactor* and its standard deviation, across all pairs of routers within the east-coast POP (sfactor=59.1% with stdev=11.8%), the west-coast POP (sfactor=64.9% with stdev=17.6%), and between pairs of routers in different POPs (sfactor=50.7% with stdev=14.3%). Overall, despite the limited aggregation benefit of using $/24$ prefixes, working sets of routers within a POP tend to be quite similar, with an sfactor of $59 - 65\%$ on average. Working sets of routers in different POPs tend to differ more, with only a 50.7% overlap on average. Some of these differences were due to localized DNS redirection (e.g., large content distribution sites redirecting users to geographically-closer servers).

4   Evaluation of Route Caching

LRU vs. LFU: In this section, we explore performance of caching algorithms directly on network traces, and start by comparing LRU (which keeps track of when each entry was used and evicts the one used the longest time ago) and LFU (which keeps track of how many times each entry is used and evicts the one used the smallest number of times while resident in the cache). Figure 2a shows that LRU outperforms LFU by a

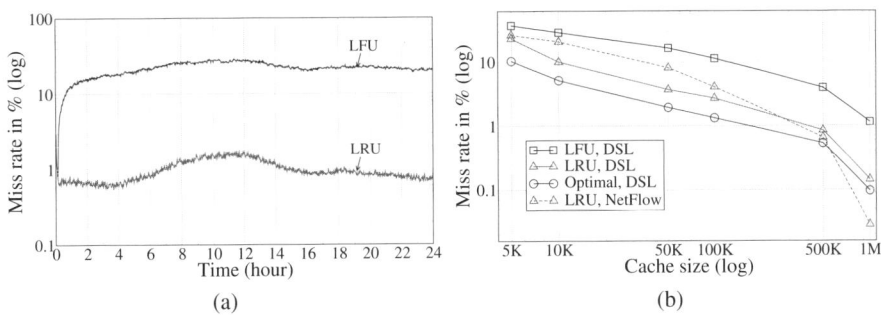

Fig. 2. Miss rate with the DSL traces, (a) Time series of miss rate, cache size = 500K, (b) Miss rates under LRU and the optimal strategy

large margin. Interestingly, the miss rate for LFU initially decreases during cold-start, which lasts for approximately 5 to 10 minutes as the cache fills. Several minutes after the cache reaches capacity, the miss rate sharply increases. This happens because the destinations of heavy (i.e., long and fat) flows become "stuck" in the cache, and the only way of evicting those is waiting for even heavier flows to overwrite those entries. LRU's miss rate also converges quickly to its long-term average and remains stable, whereas LFU's miss rate takes a few hours to converge and varies more. This happens because LFU tends to keep unnecessary entries for a long period of time. We observed these two findings across a variety of cache sizes and different input traffic mixes. Since LRU vastly outperforms LFU, we focus only on LRU for the rest of the paper.

LRU vs. Optimal: Next, we compare LRU's performance to an *optimal* scheme that has knowledge of future access times. The optimal caching algorithm (OPT) works by evicting entries that will not be needed for the longest time in the future. Note that implementing the optimal algorithm in practice is impossible (as it requires future knowledge), whereas implementing LRU (or an approximation of it) is possible – if not highly efficient or accurate. Figure 2b compares average miss rates over the unsampled DSL trace (solid curves). In a well-provisioned network, the cache would be large enough to attain a low miss rate. In this scenario, LRU performs nearly as well as OPT. For example, with a cache size of 500K, LRU's miss rate is only 0.3%-point higher than OPT's miss rate. We also study the fine-grained convergence behavior of OPT and LRU by measuring how fast their miss rates stabilize. We find that OPT's miss rate stabilizes within roughly 120 seconds, and that LRU converges almost as quickly, stabilizing within roughly 180 seconds. Given these results, it may be possible to design cache algorithms that outperform LRU on IP traffic, but it is unlikely the performance of these schemes will be substantially greater than that of LRU.

Cache size and miss rate: Figure 2b shows cache miss rates as a function of cache size for both the DSL traces (solid) and the NetFlow traces (dotted). The NetFlow curve shows average miss rate across all routers. Here we eliminated cold-start effects by measuring miss rates values only after the cache has reached capacity. We found that, with the DSL (NetFlow) traces, LRU attains a miss rate of 2.7% (4%) for a cache size of 100K, which is roughly 28% of the FIB size in conventional Internet routers. Based on the measured miss rates, we suggest some rough guidelines for determining a cache

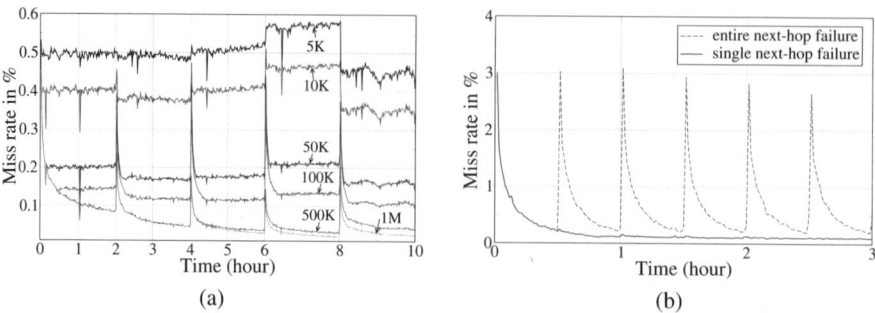

Fig. 3. Effect of routing change, (a) Inbound change (NetFlow traces), (b) Outbound change (DSL traces, cache size = 1M)

size: a cache size of 1M – which is less than $1/16$ of the maximum number of unique /24 prefixes in theory, and roughly $1/10$ of the total number of /24 prefixes used in the Internet today – may maintain a miss rate of roughly 0.1%. For a target miss rate of 1%, a cache size of roughly 500K may suffice. Overall, caching uni-class prefixes enables a route cache that is roughly an order of magnitude smaller than its full address space.

Impact of routing changes: Next, we evaluate the effect of network dynamics by manually injecting route changes into our traces, and evaluating how quickly caching algorithms converge after these changes are introduced. Two different route-change events can affect cache contents. First, routing changes upstream of the router may alter the distribution of *inbound* traffic arriving at router interfaces, leading to an abrupt change in the working set. We simulate this (Figure 3a) by randomly switching to a different router's NetFlow trace every hour while replaying against the same cache. While these events cause short-lived increases in miss rate, these spikes were roughly a factor of 3 above the average miss rate, and the miss rate stabilized in a few hundred seconds after these changes. Second, failures downstream of the router may alter the set of available *outbound* routes, causing multiple cache entries to become invalid simultaneously. We emulate (Figure 3b) the failure of multiple randomly-selected next hops by removing all cached entries associated with the given next hop upon its failure. Here, we find that for extreme scenarios where large numbers of next-hop routers fail, these spikes can be fairly large, increasing to 15 times the original value. However, for smaller numbers of failures, this value decreases substantially.

5   Related Work

Our paper is not the first to propose route caching. In fact, during the early 1990s, most Cisco routers were built with a route caching capability known as *fast switching* [2]. In these designs, packet forwarding decisions were made by invoking a user-level process that looks up a routing table (RIB) stored in slow memory. To boost lookup speeds, a route cache stored the results of recent lookups. Unfortunately, the large speed difference between the two lookup paths caused many packets to be kept in a buffer awaiting service in the slow path. In addition, upon routing changes or link failures, large groups of cached routes were simultaneously invalidated, dramatically decreasing packet forwarding rate and increasing loss probability [16]. This limitation actually led to the abandonment of route-caching and motivated the evolution of today's caching-free routers. Our findings of large bursts of consecutive misses support these earlier observations about the limitation of route caching. However, the "fall-back" scheme (explained in Section 2.1) ensures full line-rate forwarding even on cache misses by immediately sending traffic to an intermediary. Several recent works suggest that constructing a reliable and efficient traffic indirection system is possible [4, 5, 7] and thus warrant revisiting route caching.

Also there has been research which recognized the difficulty of ensuring forwarding correctness when caching CIDR prefixes [12]. These approaches increase cache size and require a logarithmic searching algorithm even when prefixes are not overlapping. Recently, Iannone et al. studied the cost of route caching under the IETF LISP architecture [7] using traffic traces to and from a campus network [17]. Hence, their analysis results are applicable to estimating caching behavior at a stub network's egress router, whereas our results are suitable to understand caching performance in a large ISP's network, where route caching would be most beneficial. Moreover, although their study was based on caching CIDR prefixes, they did not address the problem of ensuring forwarding correctness with a subset of CIDR prefixes.

To understand how *modern* workloads change the performance of route caching, we compared our results with those of earlier studies. For example, in 1988, Feldmeier studied performance of caching /24 prefixes on traces captured at a gateway connected to the ARPANET [13]. Partridge repeated a similar analysis to Feldmeier's in 1995 [18] and confirmed Feldmeier's earlier findings. We compared our results against Feldmeier's to better understand how characteristics of Internet traffic have changed for the past 20 years. By comparing the cache size needed for a target hit rate with the number of unique /24 prefixes seen in the trace, we observed some interesting results. For example, when targeting a high hit rate (larger than 98%), route caching on modern traces performs better than in these earlier studies; achieving a hit rate of 98% today requires a cache size, normalized by the number of entire /24 prefixes in the traces, of 0.1 (i.e., 10%), whereas Feldmeier reported a normalized cache size of 0.23. Moreover, when targeting a lower hit rate than 98%, modern workloads are even more amenable to caching than 20 years ago. In particular, *when targeting a sub-95% hit rate, route caching today is an order of magnitude more efficient than it was 20 years ago.* For example, for a hit rate of 95%, we found modern workloads required a normalized cache size of only 0.008, while Feldmeier reported 0.096. Traditionally a sub-95% hit rate was not considered to be tolerable, but recent routing architectures that leverage the "fall-back" mechanism can easily tolerate such a rate.

6   Conclusion

An increasing number of network architectures make use of *route caching* to achieve scalability. Evaluating the feasibility of these techniques requires rigorous evaluation of the benefits and costs of caching. This paper revisits several earlier works from the late 1980s on route caching and evaluates the practicality of their techniques on modern workloads. To the best of our knowledge, this paper constitutes the first measurement study of route-caching performance in a large ISP network. Key observations from our study are: (*i*) Working set sizes are stable over time, allowing route caches to be provisioned with relatively little headroom; (*ii*) Working sets of routers in a single POP are very similar to one another, introducing the possibility of pre-populating a cache; (*iii*) Uni-class caching eliminates complexity of longest-prefix matching and enables a cache using slower, cheaper memory; and (*iv*) Ensuring full line-rate forwarding upon cache misses is critical for the success of route caching. For future work, we plan to investigate theoretical models for the effects of sampling on estimating cache-miss rate.

References

1. Chang, E., Lu, B., Markhovsky, F.: RLDRAMs vs. CAMs/SRAMs: Part 1, http://www.commsdesign.com/design_corner/OEG20030603S0007
2. How to Choose the Best Router Switching Path for Your Network. Cisco Systems (August 2005), http://www.cisco.com/warp/public/105/20.pdf
3. Partridge, C., Carvey, P., et al.: A 50-Gb/s IP router. IEEE/ACM Trans. Networking (1998)
4. Ballani, H., Francis, P., Cao, T., Wang, J.: Making Routers Last Longer with ViAggre. In: Proc. NSDI (April 2009) (to appear)
5. Kim, C., Caesar, M., Rexford, J.: Floodless in SEATTLE: A Scalable Ethernet Architecture for Large Enterprises. In: Proc. SIGCOMM (August 2008)
6. Caesar, M., Condie, T., Kannan, J., Lakshminarayanan, K., Stoica, I.: ROFL: Routing on Flat Labels. In: Proc. ACM SIGCOMM (September 2006)
7. Farinacci, D., Fuller, V., Oran, D., Meyer, D., Brim, S.: Locator/ID Separation Protocol (LISP). Internet-Draft (work in progress) (December 2008)
8. Andersen, D., Balakrishnan, H., Feamster, N., et al.: Accountable Internet Protocol (AIP). In: Proc. ACM SIGCOMM (2008)
9. Chang, D., Govindan, R., Heidemann, J.: An empirical study of router response to large BGP routing table load. In: Proc. Internet Measurement Workshop (2002)
10. Eatherton, W., Varghese, G., Dittia, Z.: Tree bitmap: Hardware/Software IP Lookups with Incremental Updates. In: ACM Computer Communication Review (2004)
11. Sampled NetFlow, Cisco Systems, http://www.cisco.com/en/US/docs/ios/12_0s/feature/guide/12s_sanf.html
12. Liu, H.: Routing Prefix Caching in Network Processor Design. In: Proc. International Conference on Computer Communications and Networks (October 2001)
13. Feldmeier, D.: Improving Gateway Performance With a Routing-table Cache. In: Proc. IEEE INFOCOM (1988)
14. Rexford, J., Wang, J., Xiao, Z., Zhang, Y.: BGP Routing Stability of Popular Destinations. In: Proc. Internet Measurement Workshop (November 2002)
15. Jain, R.: Characteristics of Destination Address Locality in Computer Networks: A Comparison of Caching Schemes. Computer Networks and ISDN 18, 243–254 (1989/1990)
16. McRobb, D.: Path and Round Trip Time Measurements (slides 19-21), http://www.caida.org/publications/presentations/nanog9806/index.html
17. Iannone, L., Bonaventure, O.: On the Cost of Caching Locator/ID Mappings. In: Proc. ACM CoNEXT (December 2007)
18. Partridge, C.: Locality and Route Caches (1996), http://www.caida.org/workshops/isma/9602/positions/partridge.html

Quantifying the Extent of IPv6 Deployment

Elliott Karpilovsky[1], Alexandre Gerber[2], Dan Pei[2], Jennifer Rexford[1],
and Aman Shaikh[2]

[1] Princeton University
[2] AT&T Labs – Research

Abstract. Our understanding of IPv6 deployment is surprisingly limited. In fact,
it is not even clear *how* we should quantify IPv6 deployment. In this paper, we
collect and analyze a variety of data to characterize the penetration of IPv6. We
show that each analysis leads to somewhat different conclusions. For example:
registry data shows IPv6 address allocations are growing rapidly, yet BGP table
dumps indicate many addresses are either never announced or announced long
after allocation; Netflow records from a tier-1 ISP show growth in native IPv6
traffic, but deeper analysis reveals most of the traffic is DNS queries and ICMP
packets; a more detailed inspection of tunneled IPv6 traffic uncovers many packets exchanged between IPv4-speaking hosts (e.g., to traverse NAT boxes). Overall, our study suggests that from our vantage points, current IPv6 deployment
appears somewhat experimental, and that the growth of IPv6 allocations, routing
announcements, and traffic volume probably indicate more operators and users
are preparing themselves for the transition to IPv6.

1 Introduction

IPv4, the current Internet protocol, is showing its age. Addresses are becoming scarce,
with estimates of exhaustion within the next several years [1]. People are looking toward
IPv6, with its 2^{128} possible addresses, as the solution. While there has been pressure
to deploy IPv6, NAT technologies have extended IPv4's life. Given the lack of urgency
to upgrade, coupled with the administrative and financial overhead of becoming IPv6-
enabled, it is difficult to say whether we have moved any closer to a day when IPv6 is
so dominant that IPv4 can be "switched off."

Not only has IPv6 deployment has been slower than expected, but our understanding
of it is surprisingly limited as well. Questions such as, "are organizations actually using
IPv6," "what IPv6 transitional technologies are being used," and, "what applications are
using IPv6" remain largely unanswered.

To answer these questions, we looked at a variety of data sources, ranging from
regional Internet registry (RIR) logs to BGP dumps to Netflow records to packet header
traces. Along the way, we found several "gotchas," where the surface level analysis
implies a different conclusion from the in-depth analysis. For example:

- RIR data indicates that IPv6 prefixes are being allocated at near exponential rates,
  implying strong growth. However, longitudinal BGP analysis shows that nearly half
  of these allocated prefixes are never announced, and the remainder take an average
  of 173 days to appear in the global routing system. In other words, many people are
  acting as IPv6 speculators but not deployers.

S.B. Moon et al. (Eds.): PAM 2009, LNCS 5448, pp. 13–22, 2009.

– Native IPv6 traffic analysis of the enterprise customers of a US tier-1 ISP shows considerable volume, yet most of the traffic is generated by DNS and ICMP; this indicates a dearth of real IPv6 applications.
– A reasonable amount of tunneled IPv6 traffic is already observed on a US broadband ISP (0.001% of total traffic). However, further analysis indicates that much traffic is between IPv4 clients, implying that IPv6 tunneling technologies are primarily used to circumvent NAT and firewall restrictions.

The paper is organized as follows. Section 2 examines IPv6 address block allocation and compares it against an analysis of address block announcements. Section 3 looks at both native and tunneled IPv6, analyzing the types of technologies used to enable IPv6 communication and the application mix. We discuss related work in Section 4, and conclude with Section 5.

2   Allocation and Announcement of IPv6 Address Blocks

RIR and BGP data are important for understanding how IPv6 addresses are allocated and announced, respectively.

2.1   Data Sources

RIR allocations are important because they indicate the number of institutions requesting blocks, as well the sizes being allocated. For our analysis of IPv6 allocations, we used the ARIN public FTP repository [2] that maintains information about the five regional registries responsible for allocating IPv6 address blocks: APNIC, RIPE, ARIN, LACNIC, and AFRINIC. Date ranges for the different repositories are: 1999-8-13 to 2008-9-25 (APNIC); 1999-8-12 to 2008-9-26 (RIPE); 1999-8-03 to 2008-9-23 (ARIN); 2003-1-10 to 2008-9-22 (LACNIC); and 2004-12-14 to 2008-9-23 (AFRINIC).

In order to analyze how address blocks are announced, we used the RouteViews BGP data archives [3]. We collected routing table (RIB) snapshots at approximately 6 hour intervals from this web site. The BGP data obtained from RouteViews starts on 2003-5-3 and ends on 2008-9-28.

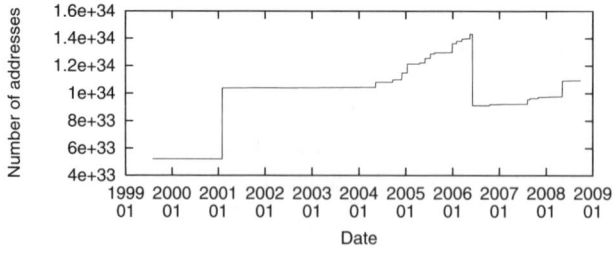

Fig. 1. Number of IPv6 addresses allocated by RIRs. Each prefix is completely de-aggregated

2.2   Why Address Allocation Statistics Are Misleading

Looking at the distribution of allocated prefixes, along with the total number of addresses allocated, seems like a reasonable method for quantifying the extent of IPv6 deployment; however, we find that using such information can be misleading.

First, one can incorrectly conclude that *IPv6 address allocations are very volatile*, as seen by the gigantic spike and dip in the curve. Figure 1 shows the total number of IPv6 addresses assigned by the RIRs. We count the number of allocated addresses by de-aggregating all prefixes into their corresponding sets of IPv6 addresses, and unioning these sets together. A couple points clearly stand out. On 2001-2-1, the number of addresses doubles, due to the allocation of $2002:/16$ to the 6to4 transitional technology [4], which reserved a part of the IPv6 space for 6to4; since this is a reservation, it cannot be considered a true measure of IPv6 growth. Likewise, a gigantic drop is seen on 2006-6-6, due to the decommissioning of the 6Bone ($3FFE::/16$). Since the 6Bone was experimental and decommissioned once IPv6 growth was observed [5], it cannot be considered evidence of significant IPv6 constriction.

Second, one can incorrectly conclude that *IPv6 growth has plateaued*. The number of allocated addresses only grew by 20% from 2006-6-7 to 2008-9-26, nearly a 2.5 year period. However, growth is masked by a few extremely large prefixes, that hide the allocation of smaller ones. As of 2008-9-28, of the 2774 allocated prefixes, 31 had a prefix length of 22 or shorter, compared to a median prefix length of 32. Since such large address blocks are allocations (*i.e.*, delegating responsibility of address assignment to local or national registries) as opposed to assignments, they are not true measures of growth. This delegation is the explanation for the plateau, and it is incorrect to draw conclusions about IPv6 growth based on it.

2.3   Drawing Correct Conclusions from the RIR Allocation

What information can be gleaned from the RIR data? After further analysis, we find that statistics concerning the number of prefix allocations provide insight into the deployment of IPv6.

Why Analyzing Prefix Allocations is Better. Since looking at the number of allocated *addresses* is not particularly insightful, why would looking at the number of allocated *prefixes* be appropriate? Moreover, is it really fair to look at prefixes independent of their size, lumping the /64s with the /48s and /32s, treating them as all "equal"?

Table 1 shows the distribution of various prefix lengths as a function of year. As time passes, more and more /32 address blocks are present, and the statistics indicate that it is the favorite for recent allocations. In fact, as of 2008-9-28, more than 67% of all prefixes allocated were /32. We believe this heavy bias is due to RIR policy changes in 2004, which made obtaining /32s easier [6,7,8].

Since so many prefixes are the same size, analyzing allocated prefixes will be roughly fair (since most address blocks are the same size), and the results will not be skewed by the few "heavy hitters" seen before.

Unfortunately, the sub-allocations are not recorded in the RIRs. Thus, if a /32 is allocated to an organization, and that organization re-allocates it to others, only the first entry is recorded. Thus, our prefix allocation analysis can potentially underestimate the growth of IPv6.

Table 1. Distribution of prefixes over time. Numbers are cumulative over time.

year	allocations	mean	mode	1st quartile	median	3rd quartile
2005-3-3	1183	33.65	32	32	32	34
2006-3-3	1421	33.39	32	32	32	34
2007-3-3	1720	34.19	32	32	32	34
2008-3-3	2179	34.65	32	32	32	34

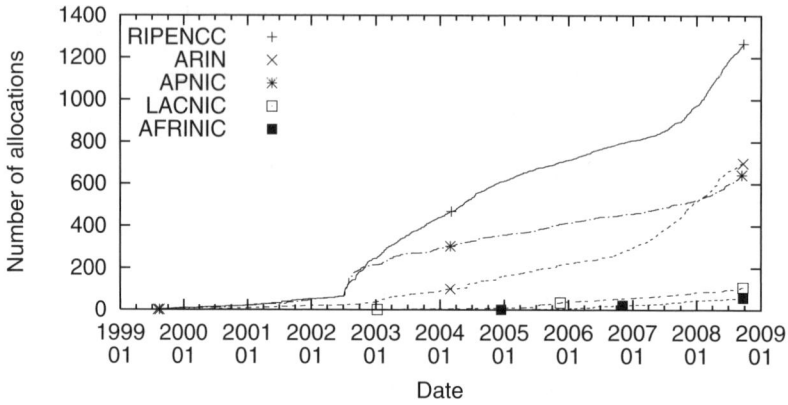

Fig. 2. Growth of the individual registries

Prefix Allocations Reveal Growth Trends. Figure 2 shows allocations of address blocks for the different registries. RIPE is clearly dominant, and shows extremely large growth for 2008. Likewise, ARIN allocations are also increasing at a very fast rate, causing it to surpass APNIC. APNIC has many allocations, but has been growing slowly, and only starts to show signs of acceleration toward the end of the study. LACNIC allocations remain approximately linear. While AFRINIC has very few allocations, it shows signs of acceleration. Cumulatively, there have been nearly 2800 address block allocations. Overall, it would appear as if IPv6 growth has been somewhat stagnant, increasing at a mostly linear rate, until recently.

One point on the graph requires explanation. In July of 2002, RIPENCC and APNIC experienced abnormal growth. Investigation revealed that on July 1st, 2002, RIPENCC and APNIC both instituted policy changes regarding IPv6 allocation [9]; this policy set new guidelines for how to allocate space, as well as guidelines for allocating to new organizations desiring IPv6 space. For example, it defined clear criteria for requesting IPv6 space, as well as a method for organizations to request additional allocations. As such, we believe that these policy changes are responsible for the sudden surge in allocations.

To summarize, IPv6 allocation has only recently started taking off; previous years had mostly linear growth, while current growth could possibly be exponential. However, since we are only at the beginning of the "knee" in the curve, we should be careful in extrapolating too far into the future.

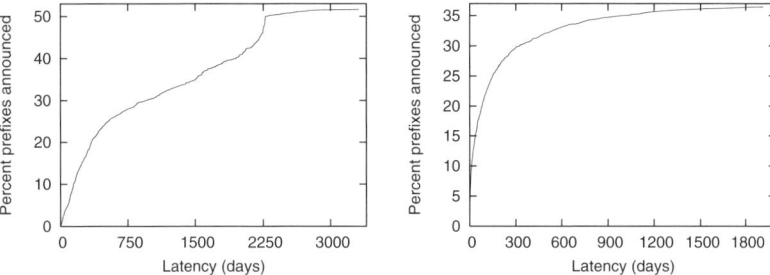

Fig. 3. CDFs of latency for prefixes where BGP data existed. Left graph represents the 52% of prefixes that were never announced. Right graph represents the 36% that were announced. The remaining 12% were allocated before our BGP data begins.

However, address allocations do not imply address usage. Is this allocation really a good measure of IPv6 deployment?

2.4 Prefixes Are Allocated, But Often Not Used

We turn to BGP routing data, which documents which IPv6 addresses are currently routable on the IPv6 Internet. Figure 3 shows how long it takes institutions to announce their address blocks (or any sub-block) after they've been allocated. We call this "usage latency," and is defined as the difference in time between allocation and the prefix's first appearance in the BGP routing tables.

There are a few points of note about Figure 3. *52% of all allocated prefixes never appear.* Measuring latency for these prefixes is impossible, since it is unknown when they will be used (if ever). Instead, we measure minimum latency, *i.e.*, the minimum amount of time before usage. We find that the average minimum latency is 957 days (which is an underestimate for the true latency!). The kink in the curve corresponds to the policy changes of APNIC and RIPENCC in July of 2002, as mentioned earlier; many addresses allocated during that surge were never used.

When computing true latency for the remaining prefixes, we run across a snag. Approximately 12% of all prefixes were allocated before 2003, when our BGP data begins. As such, it is impossible to accurately measure latency for these prefixes. We ignore such prefixes since they are a small minority.

For the remaining 36% of prefixes, latency averages 173 days. For comparison, we look to a study concerning usage latency in IPv4 in 2005 [10]. The study found that 87% of all prefixes were eventually announced in BGP, and the average latency for these prefixes was 52 days. Thus, there are fewer IPv6 users than their IPv4 counterpart, and they are also much slower to deploy.

Overall, there was some slight variation in latency between regions, but all regions were within a factor of 1.5 of each other. RIPE averaged 141 day latency. LACNIC's latency was 159 days, and AFRINIC's was 177 days. APNIC and ARIN were nearly identical, at 202 and 211 days, respectively.

3   Traffic Analysis in a US Tier-1 ISP

While the RIR and BGP data capture the rate of IPv6 adoption, this ignores three other aspects of IPv6 deployment – how people are transitioning their IPv4 networks to IPv6, how IPv6 addresses are actually assigned to individual machines, and what IPv6 applications are being used. Identifying transitional technologies helps us understand how IPv4 networks connect to the IPv6 world; The upper 64-bits of an IPv6 address identify such mechanisms, as they each have different IP ranges. Observing addresses tells us how organizations assign IPv6 addresses to individual interfaces. Since the low order 64 bits are reserved entirely for host machines, we can use this to see how individual organizations number their devices. Third, to analyze the application mix, we look at the signature *(source port, destination port, protocol)* to map it to an application name.

3.1   Data Sources

To analyze native IPv6 traffic, we use Netflow records collected from an IPv6 Internet gateway router in a US tier-1 ISP with 11 IPv6 BGP neighbors. These records were collected from 2008-4-1 to 2008-9-26, and are taken from the business customers. To analyze tunneled traffic, we collected packet header traces from 2008-7-02 to 2008-8-31 at an access router servicing approximately 20,000 DSL subscribers (different from the business customers) in an ISP. In particular, we analyzed the IPv6 headers within Teredo tunnels. Teredo [11] is an IPv6 tunneling technology created by Microsoft to enable IPv6 communications for Windows users. Due to the prevalence of Windows among typical Internet users, we assume that most tunneled IPv6 traffic destined for these subscribers use Teredo.

Unfortunately, our records are not perfect, and have some holes. Note that 5th, 6th, 9th, 10th, and 11th of July are not analyzed for Netflow. Also note that for the tunneled traffic, data from 2008-7-10 to 2008-7-21, along with 2008-8-19, 2008-8-23, and 2008-8-26 are not included.

3.2   Identifying Transitional Technologies and Address Enumeration

We identify transitional technologies as follows. Teredo uses 2001:0000: for the first 32 bits [11] of an IPv6 address, making it easily identifiable. 6to4, another popular encapsulating scheme, begin with 2002:. Although other transitional schemes exist and can be identified (*e.g.*, ISATAP, automatic tunnels, *etc.*), they are quite rare in practice; as such, we lump them together as under the label "other".

To discover how organizations assign addresses to devices, we use the same methodology as presented in [12]. The types of enumeration are: Teredo (Teredo encodes options and routing information in the lower 64-bits), MAC address based (also called auto-configuration), low (only using the last 12 bits), wordy (using words that can be spelled in hexadecimal, like BEEF), privacy (all bits are randomly set, according to the IPv6 privacy specification [13]), v4 based (when a corresponding IPv4 address influences the choice of host address), and unidentified (for all others).

Table 2. Monthly averages of different IPv6 technologies seen

(a) Native IPv6 Records

name	2008-4	2008-5	2008-6	2008-7	2008-8	2008-9
Native IPv6	85.5%	90.5%	87.0%	74.2%	63.5%	75.2%
6to4	12.7%	7.1%	10.6%	23.4%	32.4%	20.8%
Teredo	1.7%	2.3%	2.4%	2.3%	4.1%	4.0%
Other	0.1%	0.1%	0.0%	0.0%	0.0%	0.0%

(b) Tunneled IPv6 Headers

name	2008-7	2008-8
Native IPv6	70.2%	44.2%
6to4	2.8%	4.8%
Teredo	26.7%	50.3%
Other	0.3%	0.7%

3.3   Transitional Technologies

The results of analyzing the IP address structure are presented in Table 2. Most of the native IPv6 addresses of the tier-1 ISP tended to communicate with other native IPv6 addresses; approximately 80% of addresses fell into this category. 6to4 addresses were also significant, representing approximately 18% of addresses seen. Teredo addresses constituted approximately 2%, and the remaining technologies were almost negligible. These results also match those found for an analysis done in 2007 [12]. As an important note, the data sets used in our analysis are quite different from those in [12] (which included web server traffic, name server traffic, and traceroutes). Since we have a different vantage point and a different time-frame, yet have the same results, we believe that the technologies used by organizations remain unchanged for the past year.

From the tunneled perspective, we see that Teredo and native addresses are popular. Moreover, around 2008-8, a surge of Teredo-to-Teredo connections is seen.

3.4   Assigning Addresses to Machines

In addition to looking at transitional technologies, we looked at the breakdown of IPv6 address assignment schemes. Table 3 demonstrates the ratios of various host configurations. A few interesting trends emerge. First, IPv4 based addresses decline sharply (although there is a spike in August that remains unexplained). Moreover, privacy extensions remain relatively unused, occupying a small percentage of all addresses (possibly because some operating systems do not enable privacy extensions by default).

Table 3. Monthly averages of assignment schemes seen for the native IPv6 records

name	2008-4	2008-5	2008-6	2008-7	2008-8	2008-9
IPv4 Based	49.5%	28.7%	19.3%	5.9%	20.2%	6.1%
Low	22.0%	29.9%	32.5%	36.5%	31.0%	34.8%
Auto-configured	18.6%	29.2%	33.5%	40.3%	31.2%	42.6%
Teredo	1.7%	2.3%	2.4%	2.3%	4.1%	4.0%
Wordy	0.2%	0.2%	0.4%	0.3%	0.8%	0.3%
Privacy	1.0%	1.0%	1.3%	1.7%	1.5%	1.5%
Other	7.0%	8.6%	10.5%	11.8%	11.2%	10.7%

3.5 Application Mix

Looking at the application breakdown yielded interesting results, as seen in Table 4. Expected traffic, like web and mail, was surprisingly low – usually between 1% to 8% for web and 1% and 2% for mail. We performed DNS reverse lookups on the few IPv6 addresses that used web protocols and found that popular sites include an IPv6 deployment and tunnel broker and a backbone network for universities. On average, about 85% of traffic is DNS queries and 8% ICMP messages. Overall, these results are quite surprising. We believe there are two possible reasons. One could be that people are mainly using probing applications over their IPv6 networks, and not actual applications. Another is that operating systems like Windows Vista will send an extra DNS request when IPv6 capabilities are turned on: one requesting the IPv4 address and one requesting the IPv6 address [14]. Thus, the IPv6 interface may send and receive DNS queries but not traffic. Despite the potential inflation of DNS records in our data, there is still very little "real" traffic seen for IPv6. We believe that this demonstrates, for at least this tier-1 ISP, customers view IPv6 as experimental.

Table 4. Monthly averages of applications; percentages based on number of bytes

(a) Native IPv6 Records

name	2008-4	2008-5	2008-6	2008-7	2008-8	2008-9
DNS/Domain	75.5%	86.0%	87.9%	85.3%	88.8%	93.1%
ICMP	11.0%	10.2%	6.9%	6.5%	7.3%	5.2%
Web	8.3%	1.9%	1.3%	2.7%	0.8%	0.4%
Mail	1.3%	0.4%	1.0%	1.5%	0.4%	0.3%
Other	3.9%	1.5%	2.9%	4.1%	2.7%	1.0%

(b) Tunneled IPv6 Headers

name	2008-7	2008-8
Random TCP	30.6%	94.7%
Random UDP	67.3%	3.2%
Web	0.2%	0.03%
Other	1.9%	2.07%

For Teredo tunneled traffic, application breakdown was also interesting. Table 4 shows that almost all traffic is unidentifiable UDP or TCP, indicating random port numbers. Given the vast quantity of unidentifiable traffic, and the rise of Teredo pairs, it is likely that these are P2P applications communicating with each other (as random port numbers are characteristic of P2P traffic). Indeed, some applications have turned to Teredo to solve the issue faced by end hosts that are limited by their NAT/firewall technologies when they try to initiate communications with each other; using the Teredo protocol, a client contacts a Teredo server, which acts as a broker agent between Teredo clients, aiding in NAT/firewall hole punching, as well as providing unique IPv6 addresses. Several P2P clients have implemented IPv6 support [15], such as uTorrent and Vuze (formerly Azureus); moreover, uTorrent has the ability to set up Teredo automatically [16]. To summarize, it appears as if considerable tunneled IPv6 traffic is a by-product of applications (such as P2P file-sharing) using Teredo as a mechanism to bypass local NATs and firewalls, simplifying the application developers' jobs.

4   Related Work

IPv6 topology has been investigated by CAIDA's scamper work [17], as well Hoerdt and Magoni's Network Cartographer [18]. Because we did not investigate this aspect of IPv6 deployment, we consider our work to be complementary to these studies.

Anomalous BGP behavior has been analyzed through Huston's automatically generated IPv6 reports [19]. These reports include information about routing instability, prefix aggregation, table sizes, and allocation sizes.

Testing the readiness of IPv6-enabled software occurred in February of 2008, when NANOG shut off IPv4 access from their meeting for one hour [20]. It resulted in a severe restriction of services, with end users often needing to re-configure their machines. It revealed that IPv6-enabling software is still somewhat user unfriendly [21]. We believe this work on *how an individual can use IPv6* to be complementary to our work on *how organizations are using IPv6*.

Regarding traffic analysis, Arbor Networks [22] found that IPv6 traffic is growing at the same rate as IPv4 traffic. Savola [23] analyzed 6to4 traffic and found much was experimental, and also noted a rise in P2P applications. Hei and Yamazaki [24] analyzed 6to4 traffic on a relay in Japan and found that TCP traffic dominated UDP, with a considerable amount of HTTP traffic (40% of total). Our work complements these studies because we analyze different data sources, and offer a new perspective by analyzing traffic from a tier-1 ISP.

Finally, David Malone's work on IPv6 addresses analyzed transitional technologies and the assignment of IPv6 addresses to machines [12]. He looked at the breakdown of types of IPv6 addresses (Teredo, 6to4, *etc.*), as well as the classification of the host part of IPv6 addresses. While we do repeat some of the same analysis (and use some of the same techniques), we believe there are key differences between our study and his. We cover broader ground by looking at more data sources: RIR allocations, BGP data, Netflow records, and packet header traces. We also perform additional analysis, such as address space allocation and latency.

5   Conclusion

While IPv6 is beginning to see larger deployments, it still has some significant barriers to overcome. IPv6 is still viewed as experimental by some, and often is deployed in counter-intuitive ways. By analyzing RIR and BGP data, it appears that many allocations are speculatory, and that autonomous systems wait significant amounts of time before actual announcement. Moreover, although IPv6 traffic is growing, our data from a US tier-1 ISP indicated that much of it is still DNS and ICMP packets, indicating a lack of true IPv6 applications from our vantage point; additionally, tunneled traffic analysis shows much of the communication is between IPv4 pairs, implying that applications like P2P file sharing are dominant.

Further work would include a longer study of these characteristics, as well as a topological study involving more end hosts. Moreover, it would be interesting to track operating system developments and their support for various transitional schemes, as well as native support, to better understand how this software shapes the future of IPv6.

References

1. IPv4 Address Report, http://www.potaroo.net/tools/ipv4/index.html
2. ARIN public stats, ftp://ftp.arin.net/pub/stats/

3. Meyer, D.: University of Oregon Route Views Archive Project, `http://archive.routeviews.org/`

4. IPv6 Global Unicast Address Assignments (May 2008), `http://www.iana.org/assignments/ipv6-unicast-address-assignments`

5. Hinden, B.: 6bone Phaseout Planning (March 2003), `http://go6.net/ipv6.pdf`

6. ARIN Number Resource Policy Manual (August 2008), `http://www.arin.net/policy/nrpm.html`

7. IPv6 Address Allocation and Assignment Policy (August 2008), `http://www.apnic.net/policy/ipv6-address-policy.html`

8. Policy - AfriNIC IPv6 Address Allocation and Assignment Policy (March 2004), `http://www.afrinic.net/docs/policies/afpol-v6200407-000.htm`

9. Roseman, B.: ASO Statement Regarding IPv6 Policy Adoption (July 2002), `http://www.icann.org/en/aso/ipv6-statement-11jul02.htm`

10. Meng, X., Xu, Z., Zhang, B., Huston, G., Lu, S., Zhang, L.: IPv4 address allocation and the BGP routing table evolution. In: SIGCOMM Computer Communication Review (2005)

11. Teredo Overview (January 2007), `http://technet.microsoft.com/en-us/library/bb457011.aspx`

12. Malone, D.: Observations of IPv6 addresses. In: Claypool, M., Uhlig, S. (eds.) PAM 2008. LNCS, vol. 4979, pp. 21–30. Springer, Heidelberg (2008)

13. Narten, T., Draves, R., Krishnan, S.: Privacy Extensions for Stateless Address Autoconfiguration in IPv6 (September 2007), `http://tools.ietf.org/html/rfc4941`

14. Domain Name System Client Behavior in Windows Vista (September 2006), `http://technet.microsoft.com/en-us/library/bb727035.aspx`

15. SixXS - IPv6 Deployment & Tunnel Broker :: IPv6 BitTorrent Clients (September 2008), `http://www.sixxs.net/tools/tracker/clients/`

16. Forum.utorrent.com / uTorrent 1.8 released (August 2008), `http://forum.utorrent.com/viewtopic.php?id=44003`

17. Huffaker, B., Claffy, K.: Caida: research: topology: as_core_network, `http://www.caida.org/research/topology/as_core_network/ipv6.xml`

18. Hoerdt, M., Magoni, D.: Distribution of multicast tree states over the IPv6 network topology. In: 2004 IEEE Conference on Communications (2004)

19. Huston, G.: IPv6 Reports, `http://bgp.potaroo.net/index-v6.html`

20. Smith, P.: IPv6 Hour at NANOG42 (January 2008), `http://www.nanog.org/mtg-0802/ipv6hour.html`

21. Doyle, J.: IPv6 Hour at NANOG: A Follow-Up (February 2008), `http://www.networkworld.com/community/node/25276`

22. Iekel-Johnson, S., Labovitz, C., McPherson, D., Ringberg, H.: Tracking the IPv6 Migration (2008), `http://www.arbornetworks.com/IPv6research`

23. Savola, P.: Observations of IPv6 Traffic on a 6to4 Relay. In: ACM SIGCOMM Computer Communication Review (January 2005)

24. Hei, Y., Yamazaki, K.: Traffic analysis and worldwide operation of open 6to4 relays for IPv6 deployment. In: 2004 Symposium on Applications and the Internet (2004)

Analyzing Router Responsiveness to Active Measurement Probes

Mehmet H. Gunes[1] and Kamil Sarac[2]

[1] University of Nevada - Reno, Reno, NV 89557
mgunes@cse.unr.edu
[2] University of Texas at Dallas, Richardson, TX 75080
ksarac@utdallas.edu

Abstract. Active probing has increasingly been used to collect information about the topological and functional characteristics of the Internet. Given the need for active probing and the lack of a widely accepted mechanism to minimize the overhead of such probes, the traffic and processing overhead introduced on the routers are believed to become an important issue for network operators. In this paper, we conduct an experimental study to understand the responsiveness of routers to active probing both from a historical perspective and current practices. One main finding is that network operators are increasingly configuring their devices not to respond to active direct probes. In addition, ICMP based probes seem to elicit most responses and UDP based probes elicit the least.

Keywords: Internet measurement, active measurements.

1 Introduction

Internet has become one of the largest man made systems with a significant impact in many aspects of our daily life. Due to the tremendous growth in its size and importance, many groups, organizations, and governments have become interested in understanding various characteristics of the Internet for commercial, social, and technical reasons. In general, Internet measurement studies can be divided into two as (1) active measurement and (2) passive measurement studies. Active measurement studies can also be divided into two as (1) the ones that require participation from the network devices (i.e., routers) and (2) the ones that involve end systems only.

Active measurement studies that require router participation typically send measurement probes to routers and expect to receive responses from them. Naturally, such probes incur processing and traffic overhead on the routers. Upon receiving a probe message, a router is expected to create response packet and send it back to the probe originator. Most routers perform this processing on the slow forwarding path. This introduces a significant processing overhead as compared to simple packet forwarding on the fast forwarding path at the routers. In addition to network measurement studies, today many popular overlay and peer-to-peer network applications utilize active measurements to optimize their topology and/or routing performance.

S.B. Moon et al. (Eds.): PAM 2009, LNCS 5448, pp. 23–32, 2009.

In this paper, we conduct an investigation on the responsiveness of routers to active network measurements. Our goal is to quantify the responsiveness to measurement activities in two directions (1) historical perspective in terms of router participation in supporting active measurements and (2) today's best practices employed by network providers in allowing different types of network measurements on their routers. For the historical study, we use router anonymity as observed in traceroute outputs and for today's best practices we use different types of active probes and observe the responsiveness of routers to them.

For the historical study, we use path traces collected by skitter [11] and utilize one trace set for each year starting 1999. We look at the ratio of occurrences of '*'s in each path trace in data sets and use these occurrences to indicate router anonymity, i.e., lack of router participation to active probing. We study the data set both before and after processing the raw data to build the corresponding sample topology map (i.e., before and after resolving IP aliases and anonymous routers). Our results show that there has been an increase in the ratio of router anonymity in time and especially after the year of 2004. Our results also show some interesting trends on the locality of anonymous routers in path traces.

For the analysis of current best practices, we collected a set of IP addresses (536K+ of them) from a recent skitter and iPlane [10] measurement studies. These IP addresses are known to respond to indirect probes by skitter and iPlane systems. An indirect probe has a different destination than the routers that it elicit responses whereas a direct probe has a destination IP address of the intended router. In our work, we send different types of direct probe messages to each of these IP addresses and observed their responsiveness. One main observation we have is that routers are most responsive to ICMP based active probes. This is followed by TCP and then by UDP based probes. We also classified the IP addresses based on their top-level domain extensions and observed that routers in different type of organizations (e.g., .net, .com, .edu, .org, and .gov) present a similar behavior in response to direct active probes.

The rest of this paper is organized as follows. The next section presents the related work. Section 3 classifies anonymity types. Section 4 present our observations on the historical data set. Section 5 presents our observations on the current data set. Finally, Section 6 concludes the paper.

2   Related Work

Active probing has increasingly been used in various contexts to observe different characteristics of the underlying network. As an example, several research studies utilize active probing to monitor the routing and reachability behavior of the Internet in the inter-domain scale [9,11,13]. On the other hand, many overlay or peer to peer network applications depend on active probing to optimize the performance of their applications [12]. Based on the increasing need for active measurements, the research community has developed several large scale distributed measurement platforms (e.g., ark [1], Dimes [13], DipZoom [15], iPlane [10], PlanetLab [2], skitter [11], rocketfuel [14] etc.) that are commonly used to conduct various measurement activities that include active probing.

As the volume of active measurement practices increased, several researchers pointed out the impact/overhead of active probing in the network and presented approaches to reduce the volume of redundant active probes in measurement studies. Nakao et al. pointed out the increase in active measurement traffic and proposed a *routing underlay* to unify all measurement activities [12]. In this approach, overlay networks query the routing underlay for measurement information and the routing underlay extracts and aggregates topology information from the underlying network and uses this information to answer the queries.

Within the specific context of traceroute based topology measurement studies, Donnet et al. presented one of the early work on minimizing the active probing overhead in traceroute based topology discovery [4,5]. This work presented doubletree, a mechanism to avoid unnecessary probing of the same set of routers multiple times in a traceroute based topology discovery. A follow up work by Donnet et al. presented a modified approach called windowed doubletree that improves on doubletree by extracting missing links/nodes in discovering a targeted network topology [3]. More recently, Jin et al. considered scalable discovery of the underlying network topology among a group of nodes that participate in an overlay network [8]. Their main idea is to utilize a network coordinate system in identifying path traces to collect at each overlay node so as to discover the underlying topology while issuing the minimum number of path traces.

In summary, most of the related work in the area has been on reducing the unnecessary overhead in active probing based topology discovery studies. The main goal of the work presented in this paper is to quantify the responsiveness of routers to active probing based network measurement practices. The results presented in this paper provide a historical perspective and demonstrate the current practices of network operators to accommodate active network measurements.

3 Types of Router Anonymity

In our work, we measure router unresponsiveness (i.e., anonymity) to active probes. Active probes are divided into two types (1) direct probes and (2) indirect probes. A *direct probe* is the one where the destination IP address in the probe packet is the intended destination as in ICMP ping. An *indirect probe* is the one where the destination IP address in the probe packet is some other destination as in traceroute when it probes an intermediate router during the trace.

In both direct and indirect probing, eliciting a packet from probed node indicates the responsiveness of the node. On the other hand, not receiving a response to an active probe packet may have different interpretations. In the case of direct probing, the lack of a response message may not necessarily indicate node anonymity as it may be that the probed node may be unreachable; may be disconnected or turned off; or either probe or the response packet may be filtered out at some node in the network. In the case of indirect probing as in traceroute, the lack of a response message, in general, indicates node anonymity especially if another responsive node appears later on within the same trace output.

We define several anonymity types for nodes/routers that we observed in our experiments. Note that, both Type 1 and Type 2 can be further classified into two types. However, as an observer there is no difference between them.

Type 1: A router may be configured to ignore certain probe packets causing it to be anonymous with such probing. In addition, a border router may be configured to filter out (i) certain types of packets (e.g., unsolicited UDP packets directed to a local host) or (ii) outgoing ICMP responses originating from nodes within its local domain. Filtering at the border causes the internal nodes to be seen as anonymous as well.

Type 2: A router may apply ICMP rate limiting and become anonymous if the rate of the incoming probes exceed the preset limit. Similarly, a router may ignore probe packets when it is congested but may respond to them when it is not congested. In either case, the router has changing responsiveness.

Type 3: A router may have a private (publicly unroutable) IP address. Such IP addresses cannot guarantee node uniqueness and hence their originators need to be considered as anonymous.

Finally, as the above discussion suggests, the classification presented in this paper are approximate in their nature as it is difficult to know the actual cause of a lack of response for an active probe packet.

4   Historical Perspective

In this section, we use traceroute collected historical data sets to study router reaction to indirect probe messages. We downloaded 10 sets of ICMP traceroute path traces from CAIDA's skitter web site which is the only source we are aware of publicly providing periodic historical topology data. These data sets were collected within the first collection cycle in January of the each year by the skitter system. The web site reports that they had several updates to destination IP address lists. They also had a major change in their topology collection system in mid 2004 where they utilized dynamic destination lists with increased probing frequency at skitter monitors.

In the first step, we look for a trend in the ratio of anonymous routers in the collected data set. We do this before and after processing the raw path traces. Our processing in this context includes IP alias resolution and anonymous router resolution tasks. For IP alias resolution, we use an analytical inference approach called AAR [6]. Note that due to the historic nature of the data (i.e., due to the fact that the underlying topology might have changed substantially in time), a probe based IP alias resolution approach is not considered. The implication of this limitation is that the resulting topology map may have inaccuracies due to the limitations in alias resolution. Especially, data sets in 2001-2003 had much fewer aliases which we think is due to poor alias resolution. For anonymous router resolution, we use a graph-based induction (GBI) approach [7]. Table 1 presents the main characteristics of the results where "#Srcs" indicate the number of vantage points; "Reached" gives the percentage of traces that reached the final

Table 1. Analysis of historical responsiveness

Year	#Srcs	#Traces	Reached	#Nodes	Anonym	1	2	3	#Nodes	Anonym
			Initial			Anonym Type (%)			Final	
1999	5	3.5M	86.5 %	0.2 M	59.0 %	41.1	0.0	58.9	17 K	1.1 %
2000	14	14.8 M	83.5 %	0.7 M	80.6 %	19.8	0.0	80.2	18 K	1.8 %
2001	17	13.4 M	73.6 %	2.1 M	72.7 %	27.4	0.0	72.6	575 K	0.7 %
2002	20	19.1 M	50.4 %	1.5 M	51.2 %	48.9	0.0	51.1	369 K	0.8 %
2003	23	24.3 M	54.3 %	1.9 M	42.0 %	58.1	0.0	41.9	703 K	0.6 %
2004	23	22.9 M	53.0 %	2.4 M	64.1 %	36.0	0.0	64.0	45 K	1.0 %
2005	22	21,0 M	46.4 %	6.8 M	85.9 %	71.8	2.3	25.9	86 K	8.5 %
2006	19	18.4 M	37.2 %	6.4 M	87.4 %	61.3	2.2	36.5	73 K	9.7 %
2007	18	17.5 M	30.6 %	4.9 M	85.3 %	91.9	4.2	3.9	79 K	12.5 %
2008	11	10.7 M	23.2 %	2.8 M	77.2 %	90.9	7.1	2.0	61 K	9.4 %

destination; "# Nodes" gives the number of nodes within the data set before processing (i.e., before IP alias and anonymous router resolutions); "Anonym" gives the percentage of anonymous nodes in the original data set. The next three columns give the classification of anonymous nodes as percentage values. The last two columns presents final topology size and percentage of anonymous routers after processing.

According to the table, the ratio of path traces reaching their final destination decreases in time. In addition, the ratio of anonymous nodes after processing has a big increase after 2004 (see the last column). The table also shows some increase in this ratio before processing but not as much. Another observation from the column "Type 2" is that the ratio of routers employing rate limiting on active probes has increased in time. These anonymous routers had IP addresses aliased to anonymous nodes in different path traces. On the other hand, "Type 3" anonymity seem to reduce significantly during the last two years increasing the rate of "Type 1" anonymity. This might be due to recent practices of dropping incoming packets with private source addresses. We also observed that in some cases a trace source was behind a NAT and there was a high number of "Type 3" anonymity close to the probing source. Finally, for 2001-2003 data sets after processing, the "# Nodes" is substantially larger than the other years. Our IP alias resolution process returned relatively smaller number of alias pairs for these data sets. The net affect of this outcome is that most IP addresses observed in the original data set appeared as unique nodes within the final data set. This then resulted in a final topology with a large number of nodes.

In the second step, we are interested in the length distribution of path segments formed by one or more consecutive '*'s in path traces. Note that in this analysis, we do not include Type 3 anonymity in order to observe behavior of the routers to active probe messages only. We call a path segment in the form of a $(IP_1, *_1, *_2, \ldots, *_l, IP_2)$ a *-substring of length l. We are then interested in the frequency distribution of *-substrings with respect to their length l. Although a *-substring of length one may have different interpretations about the cause of router anonymity, the occurrences of *-substrings with large length values may be an indication of ISP policy in preventing active probing in its network.

Table 2. *-Substring distribution

Year	Unique *-substrings	Same AS	*-substring length				
			1	2	3	4	5
1999	1	100 %	1	-	-	-	-
2000	24	12.5 %	24	-	-	-	-
2001	57	22.8 %	57	-	-	-	-
2002	41	9.8 %	41	-	-	-	-
2003	79	20.3 %	79	-	-	-	-
2004	86	44.2 %	86	-	-	-	-
2005	225,456	12.6 %	151,133	63,662	6,360	4,301	-
2006	207,067	11.6 %	137,829	59,171	5,828	4,239	-
2007	305,331	14.4 %	212,263	73,263	14,019	5,779	7
2008	231,633	14.0 %	148,182	63,944	13,733	5,772	2

Table 2 on the right hand side presents the frequency distribution of *-substrings w.r.t. their length. In this table, we present the number of unique *-substrings in the original data set. For uniqueness, we represent a *-substring of length l as a triplet (IP_1, l, IP_2) and avoid counting the duplicate triples of this form in the table. The results in the table show a bimodal behavior where data sets prior to 2005 have only length $l = 1$ *-substrings. Starting from 2005, we observe *-substrings of larger lengths with the majority of *-substrings being of length $l = 1$ or $l = 2$. We partly attribute the bimodal behavior of the routers to changes in the skitter data collection process such as the increased probing rate and dynamically updating destination lists. Longer *-substrings might also be due to growth in networks where more hops of an autonomous system are traversed or due to increased use of MPLS. One general observation in this part is that within the last decade ISPs have become less cooperative to active probing and configure their routers to stay anonymous to such probes.

We also classify *-substrings into two based on their topological position. That is, for a given *-substring, say (IP_1, l, IP_2), we look at the relation between IP_1 and IP_2. If the IP address share the same 16-bit IP prefix, then we assume that these IP addresses belong to the same domain and therefore the anonymous nodes in between also belong to the same domain. We look at the ratio of this type of *-substrings in the data sets and present them in the "Same AS" column in Table 2. An alternative approach for this classification would be to map each such IP address (IP_1 and IP_2) to their corresponding Autonomous System (AS) number and do the check. But, since we did not have access to IP-to-AS mapping information for most of these historic data sets, we utilized the above mentioned heuristic. The results presented in the "Same AS" column suggest that most *-substrings after 2004 (e.g., more that 85%) were in the form that the two IP addresses IP_1 and IP_2 seemed to belong to different domains. The ratio prior 2005 did not show a consistent trend which we contribute to the relatively small number of occurrences of *-substrings during that time (see the "Unique *-substrings" column in the table). Given that most *-substrings are of length $l = 1$ or $l = 2$ from the right hand side of the table, we suspect that the majority of

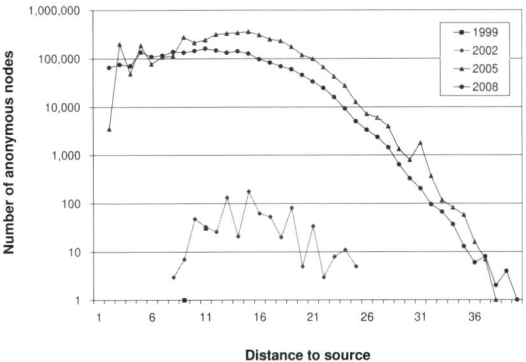

Fig. 1. Distance distribution

*-substrings that occurs in data sets after 2004 originated from routers at domain boundaries or exchange points between neighboring ASes. We also observed that even though the number of path traces used in our study decreased after 2004, the number of unique *-substrings increased during the same time period.

In the next step, we are interested in the position of anonymous nodes within path traces. In order to observe anonymous node positions in our path traces, we counted the number of anonymous nodes at each hop distance from trace sources in all path traces grouped by years. Figure 1 presents the distance distribution of anonymous nodes for four different years as samples. Similar to the previous case, this analysis excludes nodes due to Type 3 anonymity. Note that, the results presented in the figure do not consider the path length distribution of individual path traces in these data sets. According to the figure, early data sets (i.e., before 2005) contained small number of anonymous nodes that were mostly distributed 10 to 20 hops away from the source. On the other hand, recent data sets included much more anonymous nodes majority of which appeared 3 to 25 hops away from the source. The figure also shows a high number of anonymous nodes at a distance of 2 from the source for the 2008 data set. A close examination of the corresponding data set shows that this case is due to the existence of an anonymous router at a 2 hops distance to one of the vantage points.

5   Current Practices

In this section, we present our findings in an experimental study where we observe router responsiveness to direct and indirect probe messages. We use 536,743 IP addresses obtained from skitter and iPlane project web sites. These IP addresses were recently collected (between Apr 7-11, 2008) by running traceroute queries in the Internet, i.e., they belong to routers/systems that recently responded to indirect probe messages (i.e., traceroute probes issued from skitter and/or iPlane systems).

In our study, we first issued UDP, TCP, and ICMP based direct probes to each of these IP addresses and recorded the response (or the lack of it). For each

Table 3. Responsiveness to direct probes

Year	All	Router	End-Host	.net	.com	.edu	.org	.gov
# IPs	537 K	320 K	217 K	25.5 K	10.1 K	5 K	1.7 K	0.5 K
ICMP	81.9 %	84.6 %	77.9 %	92.3 %	86.4 %	88.9 %	95.5 %	92.9 %
TCP	67.3 %	70.4 %	62.8 %	76.7 %	72.6 %	83.2 %	77.3 %	83.0 %
UDP	59.9 %	64.7 %	50.3 %	63.5 %	61.7 %	57.3 %	64.4 %	62.8 %

case, we issued three probes from a host in UT-Dallas network and expected to receive at least one response to consider the probed node as responsive. In general, UDP probes are expected to return ICMP Port Unreachable messages; TCP probes (TCP SYN packets) are expected to return TCP SYNACK or TCP RST messages; and ICMP probes (Echo Requests) are expected to return ICMP Echo Reply messages. Finally, we issued another set of ICMP probes with IP Route Record option set (by using the ping command available in Linux system). We observed a very small response rate (124 responses out of 536K+ probes) and therefore excluded them from discussion.

Table 3 presents the response rate that we observed during our direct probing. The first row gives the number of IP addresses (i.e., IP addresses responding to indirect probes by skitter and iPlane systems via traceroute). The following three columns give the percentage of the IP addresses (out of the absolute numbers given on the same column above) responding to direct probes by ICMP, TCP, or UDP based probes respectively. The table also groups IP addresses as belonging to routers or end systems; and classifies them based on their top level domain extensions (only 5 of them are presented).

According to the results presented in the table, ICMP based direct probes have the highest rate of responses followed by TCP and then UDP probes. The results suggest that TCP based probes are more welcomed than the UDP ones. However, during our TCP based active probing, we received several security alert e-mails from a national ISP indicating that our probe messages were detected by their monitoring system as possible network scan activity and we were asked to stop our probing. On the other hand, our UDP or ICMP based active probing did not raise any security alert (that we know) by this or any other ISPs. These results also indicate that many routers that respond to indirect query probes do not respond to direct query probes. The ratio of such routers ranges from around 18% for ICMP probes to 40% for UDP probes. This implies that in practice network operators are tolerating indirect active probing such as traceroute more than direct active probing such as ping.

In the next step, we issued DNS queries to obtain the host names corresponding to these IP addresses. Through DNS querying, we obtained host names of around 250K IP addresses (out of 536K IPs). We then use the host name extensions to classify our IP addresses into several groups including .gov, .net, .org, .edu, and .com and look at the responsiveness of each group of nodes to active probes. Our data set had many host names with different extensions (e.g., .jp, .fr, .tr, etc.) which we did not include in the results. According to the results given in Table 3, the responsiveness ratio of routers in different types of

Table 4. Responsiveness to indirect probes

Type	Initial				Final	
	#Traces	Reached	#Nodes	Anonymous	#Nodes	Anonymous
ICMP	306 K	93.1 %	1.0 M	68.7 %	45 K	9.7 %
TCP	306 K	73.4 %	1.0 M	72.3 %	35 K	12.5 %
UDP	306 K	45.0 %	1.5 M	86.0 %	41 K	9.4 %

institution networks present similarity with each other and also has a similar
trend as the overall results presented earlier on the same table.

In the final step, we issued ICMP, TCP, and UDP based probes toward 306K
of our IP addresses to observe router responsiveness to different indirect probes.
For this, we used ICMP, TCP, and UDP based traceroute queries toward the IP
addresses. Table 4 presents the results obtained in this experiment. According
to the table, over 93% of our ICMP based probes reached their final destination
whereas these ratios for TCP and UDP based probes were around 73% and
45% respectively. These results suggest that most network operators cooperate
with ICMP based traceroute queries but more than half of the operators block
UDP based traceroute queries in their domain. In addition, according to the
"Anonymous" labeled columns in the Initial sections of the table (i.e., before IP
alias and anonymous router processing), UDP based traceroute queries cause the
highest anonymity percentage. This is also seen in the topology size difference
under the "#Nodes" labeled column before processing. However, the trend in the
results changes after processing indicating that IP alias and anonymous router
resolution processing were effective in eliminating the large number of redundant
nodes in the initial raw data. In the final data set, the differences in topology
sizes are much smaller and the anonymity ratios are somehow close to the values
reported for the historical data for 2005-2008 period in Table 1.

6   Conclusion

In this paper, we presented an experimental study on the responsiveness of
routers to active probe messages. In our historical analysis, we observed that
responsiveness reduced during the last decade. We also observed that network
operators are increasingly using rate limiting to control the impact of such active
probes in their network. Another observation from our study is that the destination reachability considerably reduced over time indicating that systems (i.e.,
routers and end hosts) are increasingly unwilling to respond to direct probes.

In the second part of our work, we observed that routers are less willing
to respond to direct active probes as compared to indirect active probes. In
addition, our active direct and indirect probing based experiments showed that
the responsiveness of routers changes with the type of the probes; ICMP based
probes having the highest response rate and UDP based ones having the lowest
response rate. Even though TCP based probes receive responses much better
than UDP based ones, this type of probes sometimes raise security alerts at
their destinations.

References

1. Archipelago Measurement Infrastructure, http://www.caida.org/projects/ark
2. Chun, B., Culler, D., Roscoe, T., Bavier, A., Peterson, L., Wawrzoniak, M., Bowman, M.: PlanetLab: an overlay testbed for broad-coverage services. SIGCOMM Comput. Commun. Rev. 33(3), 3–12 (2003)
3. Donnet, B., Huffaker, B., Friedman, T., Claffy, K.: Increasing the coverage of a cooperative internet topology discovery algorithm. In: Akyildiz, I.F., Sivakumar, R., Ekici, E., Oliveira, J.C.d., McNair, J. (eds.) NETWORKING 2007. LNCS, vol. 4479, pp. 738–748. Springer, Heidelberg (2007)
4. Donnet, B., Raoult, P., Friedman, T., Crovella, M.: Efficient algorithms for large-scale topology discovery. In: Proceedings of ACM/SIGMETRICS, pp. 327–338 (June 2005)
5. Donnet, B., Raoult, P., Friedman, T., Crovella, M.: Deployment of an algorithm for large-scale topology discovery. IEEE Journal on Selected Areas in Communications 24, 2210–2220 (2006)
6. Gunes, M., Sarac, K.: Analytical IP alias resolution. In: IEEE International Conference on Communications (ICC), Istanbul, Turkey (June 2006)
7. Gunes, M., Sarac, K.: Resolving anonymous routers in Internet topology measurement studies. In: Proceedings of IEEE INFOCOM, Phoenix, AZ, USA (April 2008)
8. Jin, X., Tu, W., Chan, S.-H.: Scalable and efficient end-to-end network topology inference. IEEE Transactions on Parallel and Distributed Systems 19(6), 837–850 (2008)
9. Katz-Bassett, E., Madhyastha, H., John, J., Krishnamurthy, A., Wetherall, D., Anderson, T.: Studying black holes in the Internet with hubble. In: Proceedings of USENIX Symposium on Networked Systems Design and Implementation, San Fransicso, CA, USA (April 2008)
10. Madhyastha, H.V., Isdal, T., Piatek, M., Dixon, C., Anderson, T., Krishnamurthy, A., Venkataramani, A.: iPlane: An information plane for distributed services. In: OSDI (November 2006)
11. McRobb, D., Claffy, K., Monk, T.: Skitter: CAIDA's macroscopic Internet topology discovery and tracking tool (1999), http://www.caida.org/tools/skitter/
12. Nakao, A., Peterson, L., Bavier, A.: A routing underlay for overlay networks. In: Proceedings of ACM SIGCOMM, Karlsruhe, Germany, pp. 11–18 (August 2003)
13. Shavitt, Y., Shir, E.: DIMES: Let the Internet measure itself. ACM SIGCOMM Computer Communication Review 35(5), 71–74 (2005)
14. Spring, N., Mahajan, R., Wetherall, D., Anderson, T.: Measuring ISP topologies using rocketfuel. IEEE/ACM Transactions on Networking 12(1), 2–16 (2004)
15. Triukose, S., Wen, Z., Derewecki, A., Rabinovich, M.: Dipzoom: An open ecosystem for network measurements. In: Proceedings of IEEE INFOCOM, Anchorage, AK, USA (May 2007)

Topology and Delay

Inferring POP-Level ISP Topology through End-to-End Delay Measurement*

Kaoru Yoshida[1], Yutaka Kikuchi[2], Masateru Yamamoto[3], Yoriko Fujii[4], Ken'ichi Nagami[5], Ikuo Nakagawa[5], and Hiroshi Esaki[1]

[1] Graduate School of Information Science and Technology,
The University of Tokyo, 7-3-1 Hongo, Bunkyo-ku, Tokyo 113-8656, Japan
[2] Kochi University of Technology
[3] Cyberlinks co.,LTD
[4] Keio University
[5] Intec Netcore, Inc.

Abstract. In this paper, we propose a new topology inference technique that aims to reveals how ISPs deploy their layer two and three networks at the POP level, without relying on ISP core network information such as router hops and domain names. This is because, even though most of previous works in this field leverage core network information to infer ISP topologies, some of our measured ISPs filter ICMP packets and do not allow us to access core network information through traceroute. And, several researchers point out that such information is not always reliable. So, to infer ISP core network topology without relying on ISP releasing information, we deploy systems to measure end-to-end communication delay between residential users, and map the collected delay and corresponding POP-by-POP paths. In our inference process, we introduce assumptions about how ISPs tend to deploy their layer one and two networks. To validate our methodology, we measure end-to-end communication delay of four nationwide ISPs between thirteen different cities in Japan and infer their POP-level topologies.

Keywords: End-to-end measurement, network tomography, communication delay, Japanese Internet.

1 Introduction

When inferring ISP topologies and identifying locations of their network elements (such as routers), researchers often rely on ISP core network information, such as the domain names of the router interfaces. Indeed many previous works (e.g., [1,2]) rely on ISP core network information to infer ISP topologies or identify network element locations, but we observe that some of measured Japanese ISPs filter ICMP packets, therefore we cannot even have access to their core network information. And also, Zhang *et al.* points out that the domain names sometimes do not represent accurate geographical locations of network elements[3].

---

* This work has been supported in part by Ministry of Internal Affairs and Communications in Japan.

S.B. Moon et al. (Eds.): PAM 2009, LNCS 5448, pp. 35–44, 2009.

Moreover, ISPs sometimes outsource designs and deployments of their layer two networks, and therefore, they do not even know where their layer two links are laid down.

In this paper, we propose a new topology inference technique that aims to reveal how ISPs deploy their layer two and three networks at the POP (Point Of Presence) level, without relying on ISP core network information. To infer such topological properties of ISP networks, we deploy systems to measure communication delay between residential users and map the collected delay and corresponding geographical POP-level paths. Our approach is based on an assumption that communication delay between users closely depends on the length of their communication path over optical fibers or copper cables. Even though it is true that a transmission delay in an access network is relatively large due to its lower speed, we could eliminate the factor under some circumstances (described in Sec. 2). By eliminating access delays, we try to map core network delays, which are derived from end-to-end delays and access delays, and their corresponding POP-level paths.

Since how carrier services lay their optical fibers is one of the important issues to map them in practice, we introduce Japan specific circumstances that major carrier services (e.g., KDDI[1] and SBTM[2]) were established as part of other infrastructure services such as railroads or expressways and optical fibers are presumably laid along those infrastructures. Through leveraging distances derived from those infrastructures, we try to map the core network delays and corresponding geographical paths. The result reveals that the Japanese Internet has the following two characteristics: 1)Some of the measured ISPs have hub-and-spoke topologies where hubs are the most populated cities in Japan such as Tokyo and Osaka; 2)All of the ISPs exchange their customer traffic at the cities.

The rest of this paper is organized as follows. In Section 2, we briefly describe our inference methodology. Section 3 shows our measurement environment and approach, in practice. In Section 4, we classify Japanese ISPs and infer POP-level ISP topologies based on a classification. We then present related works in Section 5 and finally summarize the discussion of this paper in Section 6.

2   Inference Methodology

Since our motivation of this work is to explore where ISPs deploy their POPs and how POPs are connected with each other, through end-to-end delay measurements, we briefly describe a communication path between residential users. When residential users communicate with each other, communication paths between them consist of both access (layer two) and ISP core (layer three) networks. Therefore, an end-to-end communication delay between residential users can be described as below.

$$delay(src, dst) = ad_{src} + ad_{dst} + CD(src, dst) + E_{src,dst} \tag{1}$$

---

[1] http://www.kddi.com

[2] http://www.softbanktelecom.co.jp/

Here, src and dst are nodes connected to the Internet, and communicate with each other; $delay(src, dst)$ denotes an end-to-end communication delay between src and dst; ad_{src} is the access delay at src and $CD(src, dst)$ is the delay of ISP core networks between src and dst; $E_{src,dst}$ is the measurement error of the delay.

If Internet access services are served by LECs (Local Exchange Carriers), especially in case that LECs provide DSL (Digital Subscriber Line) services, a detailed communication path can be described as follows: (1) measurement node ↔ BRAS (Broadband Remote Access Server) that aggregate user sessions from the Internet access services; (2) BRAS ↔ ISP CE (Customer's Edge) router that is located in the closest POP to users; (3) ISP core network (shown in Fig. 1). Although all the customer sessions are aggregated at BRAS not depending on which ISP users connect to, each ISP's CE routers can be deployed anywhere based on ISP policies. ISPs that serve the access services by themselves (e.g., CATVs) also deploy CE routers, which are the same as customers' default routers. So, through measuring the end-to-end communication delay and the access delays individually, we are able to derive the core delay from (1).

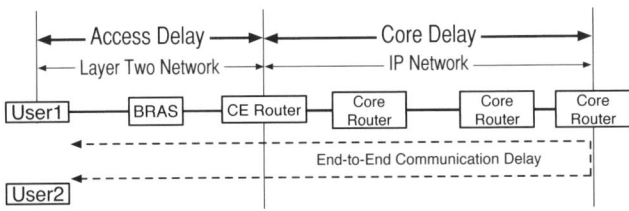

Fig. 1. Delay Model between Residential Users

To explore where ISPs deploy their POPs and how they are connected with each other, we need to map the core delay and corresponding POP-level paths. If we are able to select candidate POP locations and links among them, (1) can be transcribed into the following set of simultaneous equations.

$$delay(src, dst) = ad_{src} + ad_{dst} + \sum_{p,q \in N} x_{p,q} \times cd_{p,q} + E_{src,dst} \qquad (2)$$

Here, N denotes a set of candidate POP locations of a measured ISP; p and q satisfy $\{p, q \mid p, q \in N\}$; $cd_{p,q}$ denotes a core delay between p and q; $x_{p,q} = 1$ if a direct path between p and q exists and the path is used to connect between src and dst, otherwise $x_{p,q} = 0$. $delay(src, dst), ad_{src}$ and ad_{dst} are measurable through end-to-end measurements and $cd_{p,q}$ can be derived leveraging the distance between p and q.

In the equation, $x_{p,q} \times cd_{p,q}$ denotes the path between p and q. Since the maximum number of path patterns is almost $2^{|N|^2}$ where $|N|$ is the number of POPs, we must reduce complexities under the practical conditions of the real world. We are able to shrink the possible path patterns with the major restriction that is operations and management (OAM) cost as follows: 1) Link cost: Traffic aggregation cuts both layer one and two cost because of getting shorter optical fibers; 2) Node cost: Aggregating layer three elements (routers) reduces the maintenance costs. We will show the possible patterns in Japan according to the restriction above in Sec.4.2, which make (2) solvable.

Moreover, since it is hard to have access to layer one and two network information in general, we introduce an assumption that those networks are usually deployed along other infrastructure services, e.g. railroads and expressways, as we described in Sec. 1.

3   Measurement Environment

For a nationwide delay measurement in Japan, we deploy thirteen measurement nodes in different cities. Some of the selected cities are the most populated cities in Japan (e.g., Tokyo and Osaka), and we also choose cities that are junctions of railroads and expressways and possibly the POP locations in Japan. Each node connects to four nationwide ISPs (ISP X, Y, Z and W) through "NTT Flet's Service" that is a nationwide layer two network service for connecting end users to their ISPs via PPPoE over optical fibers.

We implement a measurement system to measure the communication delay between IP addresses attached to measurement nodes. Each node runs measurement UNIX daemons bound to four PPPoE interfaces, respectively, and each daemon measures the communication delay between the IP address and IP addresses attached to other measurement nodes. The daemon uses Linux raw socket and libpcap[3] for both sending and receiving measurement packets. The communication delay is measured by an echo request/reply method using 64-byte UDP packets. In order to minimize the queuing delay caused by network congestion, the daemon sends three train packets to each destination every ten seconds and only retains the minimum delay of them[4].

To measure the access delay, the daemon generates and send a special packet whose source and destination IP addresses are the IP address bound to the PPPoE interface. The delay measured in this manner directly corresponds to the access delay in Sec.2.

4   ISP Topology Inference

In this section, we apply our methodology to the Japanese Internet and infer POP-level ISP topologies with measured delay data. We introduce some preconditions to make our inference more accurate: 1)The velocity of light in an optical fiber becomes 60-70% compared to it in vacuum[5]. So, the velocity of light in an optical fiber cable becomes $C' = 2/3 \times C[km/sec]$; 2) There is some overhead to process the measurement packets, since we use PPPoE sessions to connect ISPs and libpcap for capturing the packets. And, there also exist queuing delays when

---

[3] http://www.tcpdump.org

the packets go through network elements such as layer two switches, layer three routers and BRAS. we define queueing delay(qd) caused by network elements as $> 1[msec]$ (this value is derived from our preliminary experiments).

4.1   Analysis of Access Delay

Figure 2 shows the minimum access delays of measured cities and ISPs in January 2008. As it shows, the access delay trends of the ISP X, Y and W are relatively similar, while the trend of the ISP Z is quite different from them. This indicates that ISP CE routers of the ISP X, Y and W are located in the almost same places. And, most of the access delays connecting to the ISPs are around $1[msec]$, we can estimate that CE routers of these ISPs are located in the same prefecture where measurement nodes are located. On the other hand, most cities except Tokyo and Osaka in the ISP Z network has long access delays that are more than 5[msec], and this implies that the ISP Z does not deploy its CE router in each prefecture. Since the access delays of the ISP Z at Tokyo and Osaka are almost the same values of the other ISPs', we can estimate that the ISP Z deploys its POP at Tokyo and Osaka, at least.

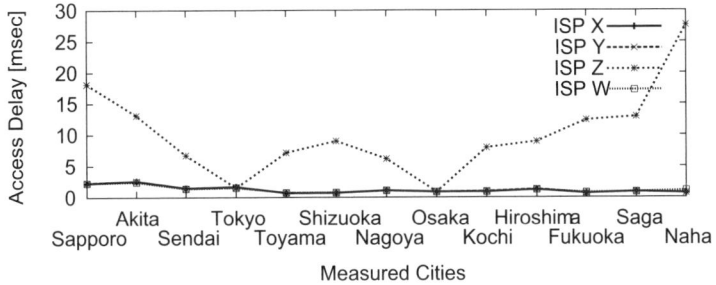

Fig. 2. Access Delay of Each City/ISP [msec]

4.2   Analysis of Core Delay

Based on the location information of the ISP CE routers in Sec.4.1, we infer ISP core network topologies. In the inference process, we use railroad distances between POPs and analyze where ISP POPs are located and how they are connected to each other. In this section, we, first, propose topology models of the Japanese Internet. Then, we try to solove (2) with the measured delay data set for the purpose of POP-level ISP topology inferences.

Topology Models of Japanese ISPs. We propose some ISP topology models of the Japanese Internet to solve (2). This intends to disclose which pairs of (p, q) make $x_{p,q} = 1$ under $E_{src,dst} < qd$ for each pair of src and dst in (2). Here, we adapt the restriction described in Sec.2 to introduce the models. And, there is one more factor that is specific to Japan. The population concentrates in Tokyo,

Type-A			Type-B	
Type-1	Type-2	Type-3	Type-1	Type-2
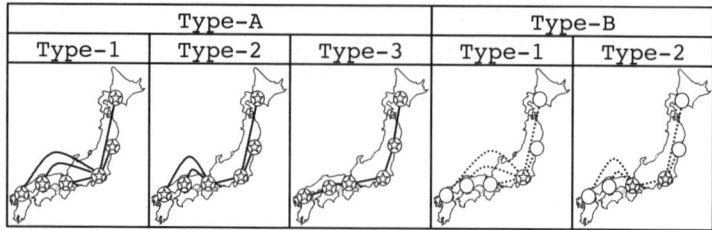				

Fig. 3. ISP Topology Classifications

Osaka, and cities along the Pacific coast between Tokyo and Osaka, therefore ISPs have planned the link and node aggregation based on these conditions.

We classify ISP topologies into the followings in Fig.3. Here, a star denotes a layer three router and solid lines between routers are an ISP backbone network. And a circle denotes a layer two switch and dashed lines between switches are an ISP access network that is same as the NTT Flet's service.

The difference between Type-1, 2 and 3 is how ISP backbone networks are structured. A Type-1 network simply aggregates all the customer traffic to Tokyo, and a Type-2 network aggregates them to Tokyo and Osaka. A Type-3 network, on the other hand, deploys routers in more or every prefecture and connects next to each other. Layer three nodes are completely reduced in a Type-1 network, and a Type-3 network maximizes link aggregation. A Type-2 network is an intermediate one of the Type-1 and Type-3. It would depend on the operation policy of a ISP's management.

The difference between Type-A and Type-B is how ISPs rely on NTT Flet's service for their layer two networks. In case of Type-A network, an ISP deploys its layer three routers in each prefecture and connects to NTT Flet's service at there. In case of Type-B networks, on the other hand, an ISP aggregates its customer traffic through NTT Flet's service and only deploys its layer three routers in the most populated prefectures such as Tokyo and Osaka. A Type-B3 network does not exist, because it indicates that all the customer traffic is exchanged through a layer two service.

Note that there are following two communication restrictions of NTT Flet's service; 1)All the PPPoE sessions connected from a measurement node are terminated at a single PPPoE accommodation called BRAS which is managed by NTT Flet's service. 2)Even though NTT Flet's service provides layer two networks for ISPs, users who connect to the same ISP at the same prefecture cannot communicate with each other through the layer two network by the nature of the service. So, if two users connect to a Type-B1 ISP, traffic between them always goes through Tokyo even if they are in the same prefecture. Since we only focus on ISPs that apply NTT Flet's service as their layer two networks in this paper, the possible network structures are covered by the above classifications. This is because NTT Flet's service should deploy layer two switches in every prefecture due to legal regulations and ISPs can only construct their network based on the layer two structure.

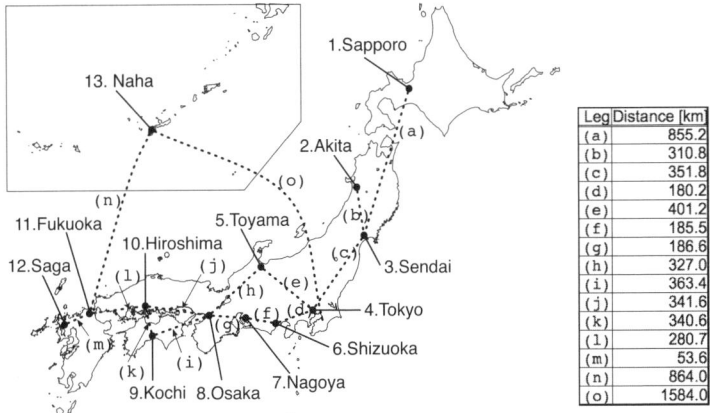

Leg	Distance [km]
(a)	855.2
(b)	310.8
(c)	351.8
(d)	180.2
(e)	401.2
(f)	185.5
(g)	186.6
(h)	327.0
(i)	363.4
(j)	341.6
(k)	340.6
(l)	280.7
(m)	53.6
(n)	864.0
(o)	1584.0

Fig. 4. Location of Measurement Nodes and Distance between Cities

Core Delay between POPs. The rest of the restrictions to solve the equations(2) is to determine the link delay denoted as $cd(p,q)$. We assume $cd(p,q) = RD(p,q)/C'$ where C' is light speed in optical fiber described in Sec.4, and $RD(p,q)$ is derived from geographical information shown in Fig.4[4]. Here, numerical symbols denote cities we set up measurement nodes and alphabetical symbols denote railroad distance RD between POPs. Since the RD is a distance between stations, there is some distance between stations and residences. We assume that the distance between users is approximately 10% longer than RD.

Inferring POP-Level ISP Topologies. Solving the simultaneous equations with the measured delays, we infer the four ISP topologies as follows.

ISP X: We use the same data set in Sec. 4.1 and Table 1 shows the end-to-end delays of the ISP X. The ISP X is a Tokyo centric network, that is a TYPE-A1 network, because the ISP deploys its CE routers in each prefecture, and the core delays between Tokyo and most of the other cities closely correlate with the distances.

One exception we find interesting is that the end-to-end delay between Tokyo and Shizuoka is larger than the corresponding delay between Tokyo and Nagoya, while the distance of the former is shorter than it of the latter. To figure out the reason, we introduce an assumption that the path between them goes through POPs away from both Tokyo and Shizuoka. The core delay between Tokyo and Shizuoka is about $8.0[msec]$, therefore the fiber length would be around $800[km]$ where the length between Tokyo and Shizuoka is about $180[km]$. Since NTT Flet's Service in Shizuoka is operated by NTT West[5] and NTT West has a huge data center in Osaka, one possible assumption is that the link detours via Osaka.

---

[4] JTB Timetable (Sep., 2008, ISBN:4910051250988)
[5] http://ntt.flets-web.com/en/west/

Table 1. End-to-End Delays of the ISP X [msec]

	1	2	3	4	5	6	7	8	9	10	11	12	13
1	2.32	36.25	27.52	23.21	35.36	31.53	27.62	29.64	36.78	34.51	37.69	39.04	52.85
2	36.37	2.63	21.08	16.92	28.39	24.50	21.40	23.19	30.73	28.34	31.36	34.38	45.70
3	27.44	20.94	1.45	9.76	20.95	16.16	14.25	16.05	20.35	21.23	24.23	32.25	37.79
4	23.21	17.23	9.85	1.58	16.99	13.14	9.88	11.67	18.78	16.63	19.74	20.81	34.32
5	34.90	27.83	20.83	16.72	0.73	22.89	21.01	24.10	27.22	27.93	31.02	39.14	44.58
6	31.43	24.62	16.03	12.58	22.87	0.77	17.93	20.97	24.05	24.70	26.04	36.01	45.41
7	27.70	21.42	14.13	9.66	21.18	17.93	1.09	16.20	23.23	20.92	24.23	25.37	39.42
8	29.48	23.30	16.00	11.62	24.08	20.57	16.13	0.94	25.30	23.11	26.21	27.50	41.89
9	36.63	30.53	20.18	18.67	27.27	24.09	23.20	25.37	0.95	30.20	30.43	34.61	45.77
10	34.57	28.27	21.15	16.52	27.92	24.75	20.93	23.17	30.19	1.25	31.07	32.32	46.36
11	37.63	31.47	24.31	19.61	31.04	26.09	24.05	26.27	30.44	31.05	0.75	37.36	47.68
12	38.99	34.37	32.37	20.95	39.12	36.08	25.36	27.62	34.65	32.37	37.52	0.98	53.10
13	52.77	45.62	37.63	33.89	44.71	45.41	39.38	42.07	45.75	46.39	47.78	53.10	0.68

So, we can conclude that the layer two network between Tokyo and Shizuoka goes through Osaka, even though Shizuoka and Osaka are not adjacent with each other in the layer three network.

ISP Y: The ISP Y has characteristics of Type-A2 network except the network aggregates traffic not only to Tokyo and Osaka but also to Fukuoka that is the largest city in Kyushu Island. We infer this as follows. Sapporo and Sendai are the neighbors of Tokyo and Akita connects to Sendai. Toyama, Shizuoka, Nagoya, Kochi and Hiroshima are the neighbors of Osaka. Tokyo, Nagoya and Osaka connect to each other. And, Saga and Naha are the neighbors of Fukuoka. Since the core delay between Tokyo and Shizuoka is larger than the expected value derived from the geographical distance, the ISP Y also has a detour path between them.

ISP Z: Since we classify the ISP Z network as a Type-B network in Sec.4.1, we only infer where the ISP Z deploys its CE routers and how each node connects to CE routers. Based on the collected delay data, we infer that the ISP Z deploys its CE routers in Tokyo and Osaka. And, Tokyo is the neighbor of Sapporo, Akita and Sendai and Osaka is the neighbor of Tokyo and Toyama, Shizuoka, Nagoya, Kochi, Hiroshima, Fukuoka, Saga and Naha. Therefore the layer three network of the ISP Z only exists between Osaka and Tokyo.

ISP W: Solving the simultaneous equation with the delay data, we infer the ISP W topology is a TYPE-A3 as follows. Most cities connect to the neighbor cities. Here, the neighbor of Toyama is Tokyo; the neighbors of Kochi are both Osaka and Hiroshima; and the neighbor of Naha is Fukuoka.

Figure 5 shows topological properties of ISP networks that we infer through the above processes. These ISP topology inferences are convinced by traceroute results and anonymous network operator sights. The results indicate that communication delay between users even in the same ISP differs depending on which ISP users connect to.

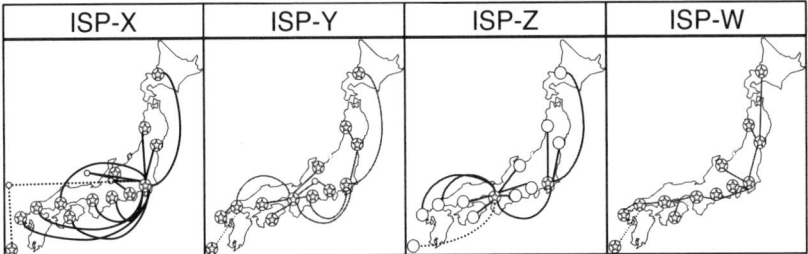

Fig. 5. ISP Topologies inferred by Our Approach

In addition, according to the result in case of ISP-X, the dissociations between layer two and three networks cannot be disclosed if we only use traceroute or other layer three information based measurements.

5   Related Work

Spring *et al.* propose Rocketfuel[1] to infer router level ISP topologies. Rocketfuel uses traceroute and aims to explore adjacency relationships between routers. Teixeira *et al.* point out that POP level topologies inferred by Rocketful have significant path diversity, and therefore they introduce the heuristic approach to improve the accuracy of the inferred Rocketfuel topologies[2]. In [6], Augustin *et al.* propose Paris traceroute to explore more accurate routing path compared to existing traceroutes. Different from these works, our approach aims to infer POP-level ISP topologies without relying on the ISP core information.

Network tomography is a research field that aims to figure out network characteristics through end-to-end measurements. Coates *et al.* introduce an overview of tomographic approaches for inferring link-level network performance[7]. Since their approach basically analyzes network characteristics from a single source point of view, Rabbat *et al.* propose new approaches that explore network characteristics through multiple source measurements[8]. We also investigate link-level network characteristics through multiple source measurements in [9] and furthermore infer ISP topologies with the same collected delay data set.

6   Conclusion

In this paper, we present a new approach for inferring POP-level ISP topologies. Since our approach leverages the end-to-end communication delay between residential users, layer one and two information and candidate POP locations for topology inferences, it is different from any previous works that require ISP core network information such as domain names of routers. Considering that ISPs tend to hide both their layer two and three structures including domain names of their equipments, our approach should become one of realistic approaches to exploring ISP topologies. Since it has been common that ISPs independently construct their layer two and three networks in these days, taking account of end-to-end communication characteristics is necessary to infer ISP topologies even for ISPs.

We apply round-trip delay measurements to infer ISP topologies based on an assumption that a round-trip path is identical. Even though this assumption is true in this paper, we need take the fact that there are asymmetric paths between users into account as our future work.

References

1. Spring, N., Mahajan, R., Wetherall, D., Anderson, T.: Measuring ISP topologies with rocketfuel. IEEE/ACM Trans. Netw. 12(1), 2–16 (2004)
2. Teixeira, R., Marzullo, K., Savage, S., Voelker, G.M.: In search of path diversity in ISP networks. In: IMC 2003: Proceedings of the 3rd ACM SIGCOMM conference on Internet measurement, pp. 313–318. ACM, New York (2003)
3. Zhang, M., Ruan, Y., Pai, V., Rexford, J.: How dns misnaming distorts internet topology mapping. In: ATEC 2006: Proceedings of the annual conference on USENIX 2006 Annual Technical Conference, Berkeley, CA, USA, USENIX Association, pp. 34–34 (2006)
4. Jacobson, V.: pathchar — a tool to infer characteristics of Internet paths, MSRI Presentation (April 1997)
5. Okamoto, K.: Fundamentals of Optical Waveguides. Academic Press, San Diego (2000)
6. Augustin, B., Cuvellier, X., Orgogozo, B., Viger, F., Friedman, T., Latapy, M., Magnien, C., Teixeira, R.: Avoiding traceroute anomalies with Paris traceroute. In: IMC 2006: Proceedings of the 6th ACM SIGCOMM conference on Internet measurement, pp. 153–158. ACM, New York (2006)
7. Coates, A., Hero, Nowak, R., Yu, B.: Internet tomography. Signal Processing Magazine 19(3), 47–65 (2002)
8. Rabbat, M., Coates, M., Nowak, R.D.: Multiple-Source Internet Tomography. IEEE Journal on Selected Areas in Communications 24(12), 2221–2234 (2006)
9. Yoshida, K., Fujii, Y., Kikuchi, Y., Yamamoto, M., Nagami, K., Nakagawa, I., Esaki, H.: A Trend Analysis of Delay and Packet Loss in Broadband Internet Environment through End Customers View. IEICE Transaction on Communications J91-B(10), 1182–1192 (2008) (in Japanese)

Triangle Inequality and Routing Policy Violations in the Internet

Cristian Lumezanu, Randy Baden, Neil Spring, and Bobby Bhattacharjee

University of Maryland
{lume,randofu,nspring,bobby}@cs.umd.edu

Abstract. Triangle inequality violations (TIVs) are the effect of packets between two nodes being routed on the longer direct path between them when a shorter detour path through an intermediary is available. TIVs are a natural, widespread and persistent consequence of Internet routing policies. By exposing opportunities to improve the delay between two nodes, TIVs can help myriad applications that seek to minimize end-to-end latency. However, sending traffic along the detour paths revealed by TIVs may influence Internet routing negatively. In this paper we study the interaction between triangle inequality violations and policy routing in the Internet. We use measured and predicted AS paths between Internet nodes to show that 25% of the detour paths exposed by TIVs are in fact available to BGP but are simply deemed "less efficient". We also compare the AS paths of detours and direct paths and find that detours use AS edges that are rarely followed by default Internet paths, while avoiding others that BGP seems to prefer. Our study is important both for understanding the various interactions that occur at the routing layer as well as their effects on applications that seek to use TIVs to minimize latency.

1   Introduction

End-to-end latencies in the Internet demonstrate triangle inequality violations (TIV). Evidence from various real world latency data sets shows that more than 5% of the triples and more than half of the pairs of nodes are part of TIVs [1, 2, 3]. TIVs are not measurement artifacts, but a natural and persistent consequence of Internet routing [4].

Triangle inequality violations expose opportunities to improve network routing by offering lower-latency one-hop *detour* [5] paths between nodes. Latency-sensitive peer-to-peer applications, such as distributed online games [6] or VOIP [7], could potentially improve their performance by exploiting TIVs [8, 9]. Consider the TIV in Figure 1, where A, B and C are all peers in the same overlay. Node A could reduce its latency to C by simply routing all traffic addressed to C through B.

Of course, all overlays violate routing policies [10]. Sending traffic along the detour paths, exposed by TIVs, instead of default paths, chosen by BGP, has the potential to disrupt traffic engineering and policy routing in the Internet. In the example above, the path ABC may violate the transit agreements between the ISPs of A, B and C (maybe because B's ISP is a customer of both A's and C's ISPs). Do all shorter detour paths violate policies? Or are they simply not selected by BGP because of its lack of mechanisms to minimize delay? Do detour paths traverse a different set of ASes that makes

S.B. Moon et al. (Eds.): PAM 2009, LNCS 5448, pp. 45–54, 2009.
© Springer-Verlag Berlin Heidelberg 2009

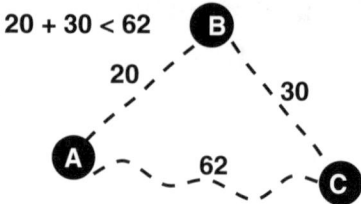

Fig. 1. Example of triangle inequality violation. All latencies are derived from real measurements.

them more attractive to users, but less attractive to ISPs? Answering such questions is important for understanding the effects of exploiting TIVs for end-to-end latency reduction.

In this paper we study the interaction between triangle inequality violations and Internet routing policies. We collect a new, large, real-world latency data set and augment it with measured and predicted AS paths. To the best of our knowledge, this is the first large (1715 nodes) latency data set that contains AS paths between the majority of the nodes. We show that, as one might expect, many of the paths of shorter detours exposed by TIVs appear impossible due to policy routing, but that 25% of them are available to BGP. Our result offers new insight into the effects of latency-reducing overlay routing as well as on how ISPs and end-users can work together to avoid less-than-optimal paths.

Our contributions can be summarized as follows:

- we present a new study on the relationship between triangle inequality violations and routing in the Internet;
- we collect a large symmetric latency data set (1715 nodes) augmented with measured and predicted AS paths; this is the first symmetric data that contains both measured RTTs and AS paths for all pairs, all collected during the same period of time;
- we show that 25% of the shorter detour paths exposed by TIVs are in fact available to BGP and could potentially provide end-to-end latency reduction without necessarily violating inter-domain policies

The rest of the paper is organized as follows. We discuss related work in Section 2. In Section 3, we present the data collection and methodology. We explore the origins of TIVs in Section 4 and discuss the relationship with BGP in Section 5. We conclude in Section 6.

2 Related Work

Previous research related to triangle inequality violations in the Internet can be grouped into two categories: studies on end-to-end latency [4, 11] and studies on the performance of network coordinate systems [1, 12, 13].

Savage *et al.* [11] measure a large number of Internet paths between geographically diverse hosts and show that alternate paths of lower latency exist between more than

20% of the pairs of nodes in their data sets. The authors study the origins of the TIVs and conclude that the availability of alternate paths does not depend on a few good or bad ASes. We confirm that no individual ASes can influence the latency of a path, but also demonstrate that the way they peer and interconnect with each other can.

Zheng *et al.* [4] use data collected between nodes in the GREN research network to argue that TIVs are not measurement artifacts, but a persistent, widespread and natural consequence of Internet routing policies. We confirm their findings that TIVs are caused by routing policies. We also study and quantify, using much larger latency and AS path data sets, the different policy decisions that may affect the formation of TIVs.

Several studies examine TIVs in relation to the impact they have on network coordinate [2, 14] and positioning [15] systems. Because these systems treat the Internet as a metric space—where TIVs are prohibited—they may obtain inaccurate results. None of these studies [1, 3, 12] considers the interaction between TIVs and Internet routing policies. Understanding the origin and the properties of TIVs would potentially help network coordinate systems to better counter the negative effects of TIVs.

3   Methodology

In this section, we describe the methodology for collecting our latency data set as well as for determining the AS paths between the nodes.

3.1   Latency Data Set

We use King [16] to compute RTTs between 1715 hosts in the Gnutella network. King uses recursive DNS queries to estimate the propagation delay between two hosts as the delay between their authoritative name servers. The IP addresses of the 1715 nodes in our measurement are provided by the Vivaldi project [2]. They were chosen such that the IPs share the same subnet with their authoritative name servers so that better-connected DNS servers would not influence the latency estimates. We run King for all pairs of IPs from a computer at University of Maryland for a week in March 2008. For each pair of nodes we keep the median of all measured latencies.

3.2   AS Paths

Understanding the AS paths beneath the TIV allows us to evaluate the detour routes for their preferences toward "better" ASes or inter-AS connections, or their compliance with known interdomain policies, *i.e.*, whether enhanced BGP protocols might find these detours and thus eliminate the TIVs. To compute as many AS paths as possible between the pairs of nodes in our latency data set we use several sources: RouteViews, Looking Glass servers and *iPlane* [17]. To the best of our knowledge this is the first large latency data set between Internet hosts augmented with AS path information computed at the same time.

RouteViews [18] collects and archives BGP routing tables and updates from commercial ISPs. We gathered AS path information from 44 BGP core routers located in 38 ISPs in March 2008. In addition, we used paths obtained by Madhyastha *et al.* [17] by probing around 25,000 BGP prefixes from 180 public Looking Glass servers.

We augment RouteViews and Looking Glass measured paths with paths predicted by *iPlane*. *iPlane* measures paths from 300 PlanetLab sites to more than 140,000 BGP prefixes to predict end-to-end paths between any pair of hosts. The predicted path combines partial segments of known paths, exploiting the observation that routes from nearby sources tend to be similar [19].

We found AS paths for the pairs of nodes in the data set, 10.4% from RouteViews and 13.6% from Looking Glass. The reason for such low completeness is that most of the Looking Glass servers and RouteViews peers are close to the core of the Internet and are unlikely to capture paths between two edge ASes. *iPlane* predicts AS paths between 71.7% of the pairs. By combining RouteViews, Looking Glass, and *iPlane*, we find AS paths for almost 75% of the pairs of nodes in the data set.

4   Origins of TIVs

To better understand TIVs and their interaction with policy routing, we must gain more insight into how they come into existence. We refer to the default path between two nodes as the direct path (or the long side of the triangle) and to the shorter path through an intermediary as the detour path (or the short sides of the triangle).

First, we look at the AS edge distribution of both direct and detour paths and show that TIVs appear not because poor ASes or AS edges are avoided by detour paths, but because detour paths are able to find better AS edges than the default direct paths. Whether these edges are known to BGP or not we discuss in the next section.

Second, we show that most latency reduction on detour paths is obtained by relaying through nodes that are either close to the source or the destination. By deviating slightly from the default path [20], one can avoid congested peering points or override routing policies that may have inflated the default path in the first place [21].

4.1   AS to AS Edge Usage

We hypothesize that triangle inequality violations appear because packets traversing the detour paths find somehow better AS edges while avoiding overloaded or circuitous edges present on the direct paths. Savage *et al.* proposed a similar hypothesis [11] on the usage of ASes on detour paths. After studying latencies between 39 traceroute servers located in North America, they showed that, for most ASes, the difference between the number of direct and detour paths in which they appeared was low. They concluded that the availability of alternate detour paths does not depend on a few good or poor ASes. We suggest that, although no individual ASes can influence the latency of a path, the way they peer and interconnect with each other can.

To study how the edges traversed by detour and direct paths are preferred or avoided, we define the weight w of an AS edge e:

$$w(e) = \frac{R(e) - D(e)}{R(e) + D(e)}, \forall \text{ edge } e$$

where $R(e)$ is the number of times e is traversed by a detour path and $D(e)$ is the number of times it is traversed by a direct path. w takes values between -1, meaning that

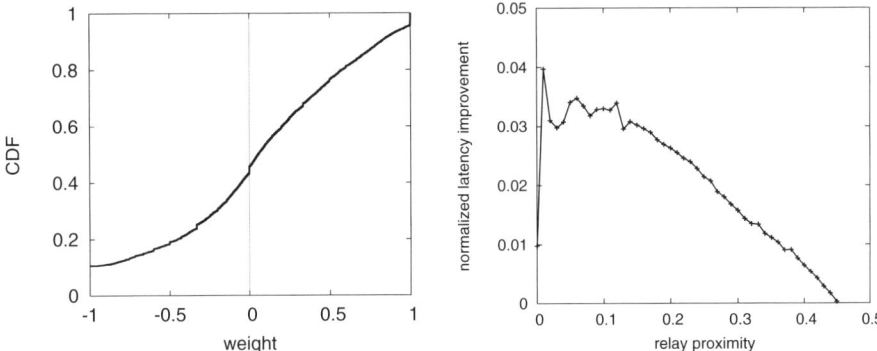

Fig. 2. (left) Distribution of weight for AS-to-AS edges: Positive values correspond to edges that are used more by detours; negative values correspond to edges that are used more by direct paths. **(right)** Distribution of normalized latencies based on the proximity of the relay to source/destination.

detours never use the edge, and 1, meaning that *direct paths* never use the edge. More generally, negative weights indicate that detours avoid the edge and positive weights indicate that detours prefer the edge.

For each pair of nodes in our data set, we find all triangle inequality violations, then select at random one *significant TIV*: a bad triangle where the detour path reduces latency over the direct path by at least 10 ms and 10%. The random selection of one bad triangle per node pair intends to not bias the results toward pathological senders or receivers, or toward the properties of the most severe violations, which might differ from the rest. We compute the weight for every edge that appears at least in one significant TIV and plot the cumulative distribution in Figure 2(left). The vertical line represents a hypothetical situation where each AS edge would be equally used by detours and direct paths. The distribution of weight is, instead, much less balanced. Detours use about 40% of all edges twice as often than the direct path does. About 10% of edges are avoided: they appear in direct paths but are never used by detours.

Thus, using preferred AS edges rather than avoiding the others is key to TIVs. One might expect that detour routing is predominantly about avoiding pathological AS paths. Our result suggests that this intuition is false: using the preferred edges is what gives detour paths lower latency.

4.2 Relay Proximity

For all TIVs in our data set, we analyze the relationship between the location of the intermediate nodes (relays) and the severity of the violation. Intuitively, if the most severe violations are obtained when relay nodes are close to either the source or the destination, then TIVs are likely due to path diversity at the endpoints.

We define the *relay proximity* of a detour path as the ratio between the latency from the relay to the closest endpoint (either the source or the destination) and the latency of the direct path associated to the detour. The relay proximity has values between 0 and 0.5; a value of 0.5 means that the relay is located at half the distance between the endpoints of the path. For each pair of nodes in our data sets, we select the detour corresponding to a significant TIV and compute its relay proximity. We group each detour according to its relay proximity and compute the total latency reduction obtained by each group. For ease of presentation, we normalize the latency reduction by dividing it to the total number of detours. We plot the results in Figure 2(right).

Most latency reduction comes when the relay is close to one of the end points, so it is likely that detours take advantage of path diversity near the endpoints.

5   Triangle Inequality Violations and BGP

It is not surprising that the BGP path selection process may prefer longer, policy-compliant paths to shorter, policy-violating detour paths. In this section we ask, to what extent are detour AS paths available to BGP?

We separate all AS detour paths in our data sets into two categories: *impossible* and *possible*. A path is *impossible* when it could not have been advertised by a neighbor, possibly because it could not have been advertised by a neighbor's neighbor and so on. Common inter-domain routing rules [22] state that customers should not advertise routes learned from a provider to peers or other providers. This prevents the customer from being used as transit between two of its providers (customer transit). Similarly, routes learned from peers are advertised to customers and not to providers or other peers, preventing peer transit. Otherwise, a path appears *possible*, though traffic engineering or other rules may have led to the selection of an alternate.

To assess whether detour paths traverse possible or impossible paths, we use the AS relationships inferred by CAIDA [23]. Directed AS edges belong to one of four categories: customer-to-provider, provider-to-customer, peer-to-peer and sibling-to-sibling. A policy compliant AS path should have zero or more customer-to-provider edges followed by zero or one peer-to-peer edges, followed by zero or more provider-to-customer edges. Sibling-to-sibling edges may appear anywhere on the path.

Table 1 classifies the detour paths. The row labeled "Unknown" corresponds to the AS paths for which we cannot give an indisputable classification using the AS relationship data set. 58% of the detour paths in the data set are non-compliant (*i.e.*, include customer or peer transit). This is not surprising, since detour paths go through end hosts, which are generally customers and may be in stub ASes. To validate our results, we performed the same classification using only the detour and direct AS paths derived from RouteViews. While the percentage of possible paths decreased only slightly (21%), we obtained more non-compliant paths (70%) and fewer unknown paths (9%). We describe the cells of Table 1 in the following discussion, first for impossible, and then for possible paths.

5.1   How Impossible Are the Impossible Paths?

We ask the following question: How severe are the policy violations of the impossible paths? For each detour path we define its prefix and its suffix. The prefix is the longest common subpath to appear at the beginning of both the detour path and a policy compliant path between the same pair of nodes, while the suffix is the longest common

Table 1. Detour paths are *possible* (may be available to the BGP decision process) or *impossible* (not advertised by BGP). Percentages inside the tables are relative to the total possible or impossible paths. Categories separated by horizontal lines overlap.

Total Detours		793,693
Impossible AS Paths		460,830 (58%)
Cause	Customer transit	343,381 (75%)
	Peer transit	117,449 (25%)
Type	Truly disjoint	302,207 (66%)
	Borderline	153,057 (33%)
	Undercover	5,503 (1%)

Possible AS Paths		197,453 (25%)
Traffic Eng.	Relay AS not on direct path	56,813 (29%)
	Direct, detour paths differ	103,215 (52%)
	Direct, detour paths same	37,425 (19%)
Path length	Shorter than direct	17,770 (9%)
	Equal to direct	75,032 (38%)
	Longer than direct	104,651 (53%)
Transit cost	Smaller than direct	35,541 (18%)
	Equal to direct	96,751 (49%)
	Greater than direct	65,161 (33%)

Unknown	135,410 (17%)

subpath to appear at their end. Based on the prefix and the suffix, we define two measures to capture the severity of policy violation of a detour path: *width* and *depth*. The width is the number of valid AS edges that would be required to connect the suffix and the prefix to obtain a policy compliant path. The depth is the minimum number of AS edges that have to be traversed from the relay to the end of the prefix or the beginning of the suffix. Based on the values of width and depth we classify the impossible detour paths into *undercover*, *borderline*, and *truly disjoint*. We present an example of each type in Figure 3 and describe them below:

undercover (depth = 0) (1% of impossible detour AS paths)
    Because the depth is 0, the relay of the detour lies on a compliant path. Although both direct and detour traffic enter the AS of the relay, they use different peering points to exit.

borderline (depth = 1, width ≤ 1) (33%)
    Borderline compliant detours diverge from the compliant path only to traverse the relay before returning quickly.

truly disjoint (all other cases) (66%)
    A truly disjoint path differs from any compliant path by at least two AS edges.

The results above show that one third of the "impossible" paths are within *one AS hop* of being "possible". Enhanced BGP protocols—where nodes exchange path

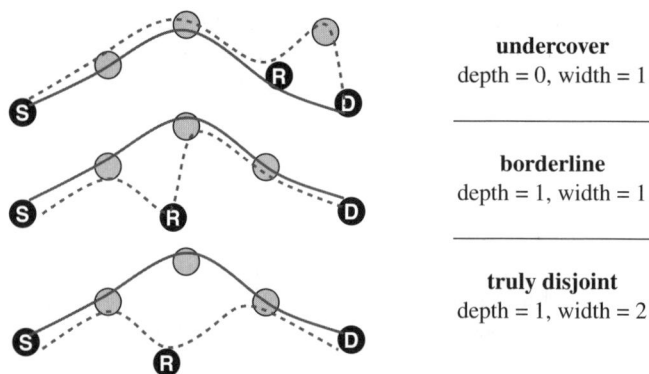

undercover
depth = 0, width = 1

_____

borderline
depth = 1, width = 1

_____

truly disjoint
depth = 1, width = 2

Fig. 3. Examples of impossible detour AS paths

performance information with their neighbors—could learn about these detours, eliminate the TIVs and offer faster paths to end users.

5.2   Possible Paths

25% of the detours in the data set follow compliant paths. Therefore, they can be learned by BGP. Only traffic engineering decisions or a lack of configuration can stop these paths from being advertised and learned. BGP routers select paths based on cost, performance, length, and even which path is advertised first. Since we do not know precisely why any path was chosen, we consider here a few possible explanations.

Traffic Engineering. Each AS must pay some cost to carry traffic in its internal network. ISPs engineer their networks and routing to minimize this cost, while improving performance, choosing early-exit routes that deliver packets at the nearest exit, or divert traffic to balance load. Although we do not have explicit information about these choices, we can infer when such traffic engineering occurs. For example, for 52% of the possible detour paths, the AS of the relay node lies *on the direct path*, yet the detour and the direct paths are different. This may occur because traffic, when redirected through the relay, will traverse a different peering point than the default traffic.

   These results suggest that detours may take advantage of shorter paths by overriding common traffic engineering practice. The number of detours due to minimizing internal cost may be higher than we have observed; we can only identify such detours when the relay is on the direct path.

Path Length. When choosing among otherwise equal paths, BGP selects the one with the fewest ASes. Because a detour path traverses an additional relay point, we expect it to use more ASes than the corresponding direct path. For each pair of nodes, we compute the difference in number of AS hops between the detour path and the corresponding direct path. Over 90% of the *possible* detour paths traverse at least as many ASes as the corresponding direct paths (Figure 4(a) and Table 1). This suggests that latency is not reduced by eliminating ASes traversed.

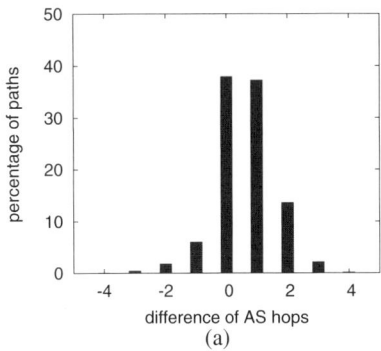 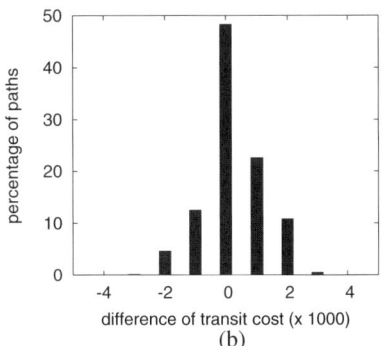

Fig. 4. Possible detour AS paths have larger (a) path length, and (b) transit cost

Transit Cost. Although not visible in BGP data, the price an ISP pays to its provider may make a path more or less preferred. Traversing larger networks implies greater expense. We define the transit cost of a path as the maximum degree—number of AS-to-AS peerings—of all ASes on the path. Table 1 and Figure 4(b) show the results of the comparison between the transit cost of detour paths and corresponding direct paths. The transit cost of the detour paths is significantly higher than that of direct paths.

6   Conclusions

In this paper, we offer new evidence into both the origins and properties of Internet triangle inequality violations. We show that triangle inequality violations occur because many AS edges are constantly avoided by BGP. By analyzing the decisions that lead to such occurrences, we show that, not surprisingly, most detour paths of TIVs violate interdomain routing policies. ISPs control Internet routing using BGP which chooses paths primarily based on cost, policies, past performance, even which route arrives first, but never based on end-to-end latency. However, we find that 25% of the paths in our data sets are available to BGP (and 20% more are borderline available): BGP knows of low latency paths but prefers them less.

Our intention is not to reprimand BGP for not being able to offer low-latency paths to its users. We show that, in fact, there is room for improvement in Internet routing, without affecting the equilibrium of the tussle [24]: both end-users and ISPs could take advantage of better paths without any of them feeling cheated. We intend to explore in future work ways in which latency-reducing overlay networks and policy routing can coexist.

References

1. Wang, G., Zhang, B., Ng, T.S.E.: Towards network triangle inequality violation aware distributed systems. In: IMC (2007)
2. Dabek, F., Cox, R., Kaashoek, F., Morris, R.: Vivaldi: a decentralized network coordinate system. In: SIGCOMM (2004)

3. Lumezanu, C., Levin, D., Spring, N.: PeerWise discovery and negotiation of faster paths. In: HotNets (2007)
4. Zheng, H., Lua, E.K., Pias, M., Griffin, T.G.: Internet routing policies and round-trip times. In: Passive and Active Measurement Workshop (2005)
5. Savage, S., Anderson, T., Aggarwal, A., Becker, D., Cardwell, N., Collins, A., Hoffman, E., Snell, J., Vahdat, A., Voelker, G., Zahorjan, J.: Detour: A case for informed Internet routing and transport. IEEE Micro. 19(1), 50–59 (1999)
6. Bharambe, A., Douceur, J.R., Lorch, J.R., Moscibroda, T., Pang, J., Seshan, S., Zhuang, X.: Donnybrook: Enabling large-scale, high-speed, peer-to-peer games. In: ACM SIGCOMM (2008)
7. Kho, W., Baset, S.A., Schulzrinne, H.: Skype relay calls: Measurements and experiments. In: IEEE Global Internet Symposium (2008)
8. Lumezanu, C., Baden, R., Levin, D., Spring, N., Bhattacharjee, B.: Symbiotic relationships in Internet routing overlays. In: NSDI (2009)
9. Andersen, D.G., Balakrishnan, H., Kaashoek, M.F., Morris, R.: Resilient overlay networks. In: SOSP (2001)
10. Qiu, L., Yang, Y.R., Zhang, Y., Shenker, S.: On selfish routing in Internet-like environments. In: ACM SIGCOMM (2003)
11. Savage, S., Collins, A., Hoffman, E., Snell, J., Anderson, T.: The end-to-end effects of Internet path selection. In: SIGCOMM (1999)
12. Lee, S., Zhang, Z.L., Sahu, S., Saha, D.: On suitability of euclidean embedding of internet hosts. In: Sigmetrics (2006)
13. Lua, E.K., Griffin, T., Pias, M., Zheng, H., Crowcroft, J.: On the accuracy of the embeddings for Internet coordinate systems. In: IMC (2005)
14. Ng, T.S.E., Zhang, H.: Predicting Internet network distance with coordinates-based approaches. In: INFOCOM (2002)
15. Wong, B., Slivkins, A., Sirer, E.G.: Meridian: A lightweight network location service without virtual coordinates. In: SIGCOMM (2005)
16. Gummadi, K., Saroiu, S., Gribble, S.: King: Estimating latency between arbitrary Internet end hosts. In: IMW (2002)
17. Madhyastha, H.V., Isdal, T., Piatek, M., Dixon, C., Anderson, T., Krishnamurthy, A., Venkataramani, A.: iPlane: An information plane for distributed services. In: USENIX OSDI (2006)
18. RouteViews: Routeviews (2008), http://www.routeviews.org
19. Madhyastha, H.V., Anderson, T., Krishnamurthy, A., Spring, N., Venkataramani, A.: A structural approach to latency prediction. In: IMC (2006)
20. Yang, X., Wetherall, D.: Source selectable path diversity via routing deflections. In: ACM SIGCOMM (2006)
21. Spring, N., Mahajan, R., Anderson, T.: Quantifying the causes of path inflation. In: ACM SIGCOMM (2002)
22. Gao, L.: On inferring autonomous system relationships in the Internet. IEEE/ACM Transactions on Networking 9(6), 733–745 (2001)
23. Dimitropoulos, X., Krioukov, D., Fomenkov, M., Huffaker, B., Hyun, Y., kc claffy, R.G.: As relationships: inference and validation. SIGCOMM CCR 37(1), 29–40 (2007)
24. Clark, D.D., Wroclawski, J., Sollins, K.R., Braden, R.: Tussles in cyberspace: Defining tomorrow's Internet. IEEE/ACM Transactions on Networking 13(3), 462–475 (2005)

Methods for Large-Scale Measurements

Queen: Estimating Packet Loss Rate between Arbitrary Internet Hosts

Y. Angela Wang[1], Cheng Huang[2], Jin Li[2], and Keith W. Ross[1]

[1] Polytechnic Institute of NYU, Brooklyn, NY 11201, USA
[2] Microsoft Research, Redmond, WA 98052, USA

Abstract. Estimate of packet-loss rates between arbitrary Internet hosts is critical for many large-scale distributed applications, including overlay routing, P2P media streaming, VoIP, and edge-server location in CDNs. iPlane has been recently proposed to estimate delay, packet-loss rates, and bandwidth between arbitrary hosts [1]. To our knowledge, iPlane is the only published technique for estimating loss rates between arbitrary Internet hosts. In this paper, we present Queen, a new methodology for estimating packet-loss rates between arbitrary hosts. Queen, extending the King [2] methodology for estimating delay, takes advantage of the open recursive DNS name servers. Queen requires neither additional infrastructure deployment nor control of the DNS recursive servers. After describing the methodology, we present an extensive measurement validation of Queen's accuracy. Our validation shows that Queen's accuracy is reasonably high and, in particular, significantly better than that of iPlane for packet-loss rate estimation.

Keywords: Recursive DNS, Retransmission Pattern, Loss Rate.

1 Introduction

End-to-end packet loss-rate and delay are fundamental network metrics, both of which impact the performance of network applications. Estimate of delay and packet-loss rate between arbitrary Internet hosts is critical for many large-scale distributed applications, including overlay routing, P2P media streaming, VoIP, and edge-server location in CDNs. Applications can measure latency and loss rate by passive monitor or active probe. For example, Akamai's EdgePlatform [3], deployed in 70 countries, continually monitors Internet traffic, trouble spots, and overall network conditions. Ensuring high-quality service to its customers, this monitoring is an indispensable component of Akamai. RON [4] measures latency and loss rate continuously, and switches overlay paths accordingly. However, active probing often requires control of at least one end of the path, which imposes a significant coverage limitation. In academia, researchers are generally limited to paths imposed by the PlanetLab platform. Due to deployment and maintenance costs, large commercial entities are limited to modest-scale deployments of measurement platforms; for example, Keynote's platform is only available in 240 locations [8]. To overcome these limitations, several schemes have been developed to provide delay estimates without access to either end of an Internet

S.B. Moon et al. (Eds.): PAM 2009, LNCS 5448, pp. 57–66, 2009.
© Springer-Verlag Berlin Heidelberg 2009

path. King [2] leverages DNS infrastructure to measure latency between arbitrary end hosts. Network coordinate systems construct virtual coordinate spaces from limited latency measurements, and then make latency predictions based on the virtual coordinates [7,9,10]. Azureus builds one of such latency estimation scheme into its production P2P client [10], in order to make better peer selection.

Although there is significant research on delay estimation between arbitrary end hosts, there has been relatively little work on packet-loss rate. To our knowledge, the only existing published methodology for estimating packet-loss rates between arbitrary end-hosts is iPlane [1], which also measures other metrics. It predicts the loss rate between arbitrary hosts by composing the performance of measured segments of Internet paths. In this paper, we present *Queen*, a new methodology for estimating packet-loss rates between arbitrary hosts. *Queen* requires neither additional infrastructure deployment nor control of end hosts.

Queen builds upon the well-known latency measure tool King [2]. *Queen* approximates packet loss rates between arbitrary hosts by finding two DNS servers near them and determining the packet loss rates between these two DNS servers. However, although *Queen* gets its initial inspiration from King, it is nevertheless very different – its design has required a deep understanding of how currently deployed DNS servers operate. In particular, we have discovered that all DNS servers are configured with highly regular retransmission mechanisms, allowing packet loss to be inferred from observed excessive latencies. Because the gap in DNS retransmissions is large and regular, *Queen* is accurate even though latencies can vary wildly between end systems over short period of time.

The contribution of this paper is as follows: *(i)*: We develop a new methodology that estimates packet loss rate between arbitrary Internet end hosts without control on either end. We first characterize the retransmission behavior of deployed DNS servers; we then propose a loss-rate formula based on this behavior. *(ii)*: Based on the methodology, we develop a tool, *Queen*, which is made public at *http://cis.poly.edu/~angelawang/projects/lossrate.htm*. *(iii)*: We conduct extensive measurements to validate the accuracy of *Queen*. In particular, we show that Queen is more accurate than iPlane for estimating packet-loss rates. *(iv)*: As a case study, we perform an Internet-wide packet-loss rate measurement. The results are informative and can also provide realistic Internet characteristics to other platforms, such as Emulab [11].

The rest of the paper is organized as follows. In Section 2, we briefly review how King works, then we present the design of *Queen* in Section 3 and evaluate its accuracy in Section 4. In Section 5, we present an Internet-wide experiment result. Afterwards, we present related work and conclusion in Section 6 and 7.

2   Brief Review of King

King, developed by Gummadi et al. [2], is a methodology to estimate latency between arbitrary Internet hosts. They propose a simple version that requires no external setup and a somewhat more complex one with much improved accuracy and additional setup. We only review the latter version for this work.

To measure the latency between arbitrary end hosts A and B, King (i) finds DNS name servers that are topologically close to A and B (say R and T, respectively); (ii) estimates the latency between the two DNS name servers R and T using DNS queries; and (iii) uses the measured latency between R and T as an estimate of the latency between A and B (see Figure 1).

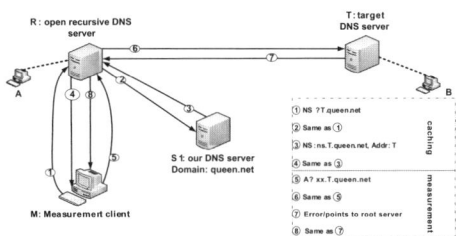

Fig. 1. King measures the latency between R and T by *tricking* R to directly query T

The key step is to estimate the latency between R and T. This requires at least one of the servers to be an *open recursive DNS server*, that is, a DNS server allowing recursive DNS queries from *arbitrary* Internet hosts. Henceforth, assume that R is such an open recursive server. The important issue is how to "trick" R to send DNS queries *directly* to T in order to measure the latency between them. The key idea is: (i) first trick R into believing that T is the authoritative name server for a special domain; and (ii) then query R for a host in this domain, which will trigger R to forward the DNS query directly to T. We refer to steps (i) and (ii) as the caching and measurement stages, respectively.

For the caching stage, we need operate an authoritative DNS server (call it S1) for a special domain (say queen.net). A measurement client M (arbitrary) sends to R a recursive NS type query containing encoded T's IP address for a sub-domain (say t.queen.net), which will ultimately be answered by S1, the authoritative DNS server for domain queen.net. S1 is programmed to reply that T is the authoritative DNS server for sub-domain t.queen.net, and this reply will be cached at R. This completes the caching stage. From this point on, any subsequent recursive query through R for a host belonging to sub-domain t.queen.net will be forwarded to T directly. Since T is actually *not* the authoritative name server for this sub-domain, it will return an error message (or a pointer to the root DNS servers). R will in turn report query failure to M. Thus, R is tricked into querying T directly, and the latency between them can be estimated.

3   Methodology

DNS queries are transmitted using UDP by default. An interesting question is: What happens if there is packets lost (either query or response) between the two representative name servers R and T? This leads us to a new methodology for estimating packet-loss rates between arbitrary Internet end hosts.

3.1 Retransmission Pattern

When a DNS server sends a query to another DNS server, either the query or the response could get lost along the path. All DNS servers have built-in mechanism to deal with such losses. In either case, if the querying name server does not receive response within a certain period of time, it will resend the query until a retry limit is reached. Intuitively, intervals between retransmissions in most DNS servers should be substantially larger than RTT of most paths. We now confirm this intuition by studying the retransmission patterns of DNS servers.

Architecture. To dig a DNS server's retransmission pattern, we force the server to resend queries until its retry limit exhausted. Fig. 2(a) shows our architecture. We operate two name servers, S1 and S2, which are configured as: 1) S1 is the authoritative name server for domain queen.net, and delegates sub-domain poly.queen.net to S2; 2) S2 runs as a simple DNS black hole, which only records the received time of all incoming queries but does *not* reply to any query. When client M sends to R a recursive DNS query for a host (say host.poly.queen.net) in the sub-domain, R will be redirected to S2 after first contacting S1. Since S2 never replies, R will resend the query until exhausting its retry limit. We encode R's IP address inside the query (together with a unique identifier), so that S2 can easily extract the address, as well as match queries. Finally, we collect the timestamps of all queries from each R, and calculate the retransmission pattern.

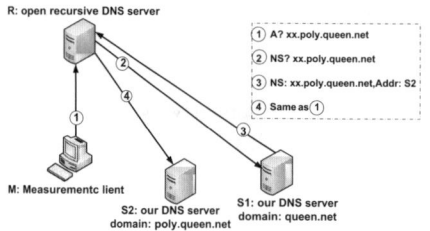

Pattern	# of Servers	% of Total
2-2-8-16	15255	51.1
2-2-8-16-30-30	7600	25.4
4-8	2106	7.0
0.5-0.5-1-2-4-8-10	1471	5.0
2-2-8	823	2.8
2-2-8-16-30	355	1.2
4-8-16	148	0.5
2-8-16	146	0.5

(a) Method to measure a name server's retransmission pattern

(b) Most Common Retransmission Patterns of DNSs

Fig. 2. Retransmission pattern measurement

Experiment. We setup an experiment to discover the retransmission patterns of ∼30,000 DNS servers picked from a large list of unique open recursive DNS servers with wide coverage, obtained in our previous study [6]. Those servers cover 6 continents and 147 countries. For each name server R, we send a unique recursive query from our measurement client M. The retry at M is set to 0 to ensure that exactly one query is sent to each R and all the duplicate queries recorded at S2 are generated solely by R. We found that there are small number of common retransmission patterns among DNS servers, showed in Fig. 2(b). Those patterns cover 93.5% measured servers. As an example, pattern 2-2-8-16 means the server will retry 4 times, 1st after 2 seconds timeout, 2nd after another 2 seconds timeout, 3rd after another 8 seconds, and 4th after another 16 seconds. Furthermore, about 94% servers will wait at least 2 seconds before retry.

3.2   Loss Rate Estimation

We are now ready to describe how to infer packet losses from large latencies, and how to estimate the packet-loss rate.

Packet Loss Definition. We will use an example to explain how to infer packet losses from large latencies. In Fig. 3, we measure the latency between two name servers R (in S. Africa) and T (in Seattle) using DNS queries with exponential inter-arrival time, where the average is 200ms and their associated counting process is Poisson, over a 15-minute duration. Here, R is an open recursive name server with retransmission pattern 2-2-8-16s. We can see most latencies fall into a range (490-500ms) – the regular RTTs between R and T. However, there are some latencies far larger than the regular ones. We infer that these large latencies correspond to packet losses. Specifically, we compute a latency threshold based on R's retransmission pattern. Because the regular RTT is about 500ms and R sends out its first retransmission after 2 seconds if there is no response, we set the threshold to be 500ms+2s. Thus, any latency around 2500ms implies one packet loss between R and T, either on forward path or reverse path. Similarly, any latency around 500ms+2s+2s implies two packet losses, and so on.

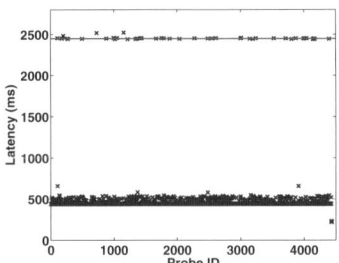

Fig. 3. Packet Loss Example

Loss Rate Computation. Now we know whether there is a packet loss from the measured latency. Next is how to compute the packet loss rate exactly. Suppose we send out N total queries, where M queries receive response and obtain the latencies, while the other $(N - M)$ queries have failed (without responses). To compute the total number of lost packets, we consider those M queries only (the reason to be explained later). With the loss definition in previous section, we can infer how many queries (say L out of M) have excessive latencies, as well as the total number of retransmission packets L' (clearly, $L' \geq L$). Then, the packet loss rate is simply calculated as $L'/(M + L')$.

So far, we have presented our methodology to estimate the packet-loss rate from excessive DNS query latencies. However, what's the effect if packets get lost between M and R, or R and S1 in Fig 1? The facts are: 1) if the packet is lost between M and R, no matter forward or reverse, M will not receive response at all. This is exactly the reason to exclude those failed queries when computing the loss rate; 2) Packet losses between R and S1 will be automatically taken care of by R's retransmission mechanism. In addition, it happens during the caching stage, so the excessive latency here will *not* affect the measurement stage at all. Thus, the losses inferred in our method will only be the losses between R and T.

Another potential problem is, when T returns an error to R, depending on its configuration, R may have two other outlier actions: 1) It may retransmit the same query several times. In this case, even without packet loss, the estimated latency will be a factor of K larger, where K is the number of retransmissions. It will introduce noise to the real loss. 2) Alternatively, it may stop forwarding further queries for the same sub-domain to T. If so, R will not contact T directly any more after an error, which causes the algorithm to fail. Fortunately, we can easily point our own DNS server as T and send trial DNS queries to examine all open recursive DNS servers (R's). By parsing the query logs on our own DNS server, we can identify which R's behave as outliers and simply filter them out in future measurements. Based on our observation, there are only few such outlier-behaving DNS servers; majority can handle exception normally.

4   Validation

In this section, we present quantitative validation of the accuracy of *Queen*.

4.1   Direct Path Validation

First, we fix the target T to be S1. We trick open recursive DNS servers (R's) to query S1 and estimate the packet loss rate between them, as shown in Fig. 4(a). We send out DNS queries with exponential inter-arrival time and 200ms in average over a 15-minute duration, so that their associated counting process is Poisson. In parallel, we also run direct probing by sending ICMP packets from S1 to R, which serves as ground truth loss rate. Note that, in this validation, Queen is estimating the packet loss rate on exactly the same path as the direct probing.

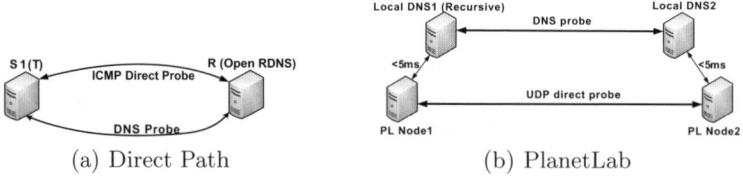

(a) Direct Path   (b) PlanetLab

Fig. 4. Validation Path Setup

We randomly choose 370 open recursive name servers from 5 continents. In the end, we get results for ~330 paths, where 210 experience loss either in direct probing or Queen. Fig. 5(a) compares two latencies. Optimally, they will align at 45° straight line. As we can see they indeed match very well. Fig. 5(b), 5(c) compare the loss rate estimated by two methods. They also match very well. In particular, the absolute loss rate difference between direct probing and Queen is

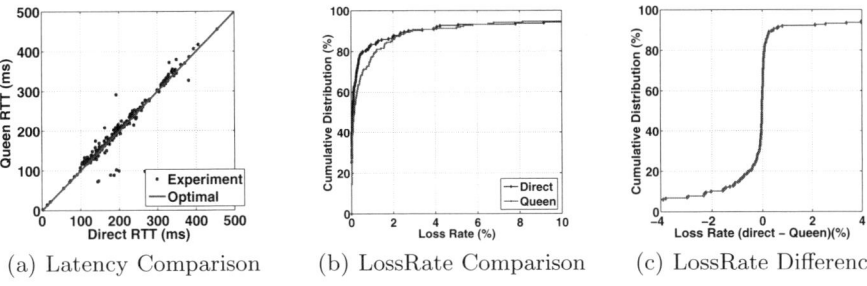

(a) Latency Comparison   (b) LossRate Comparison   (c) LossRate Difference

Fig. 5. Direct Path Validation Results

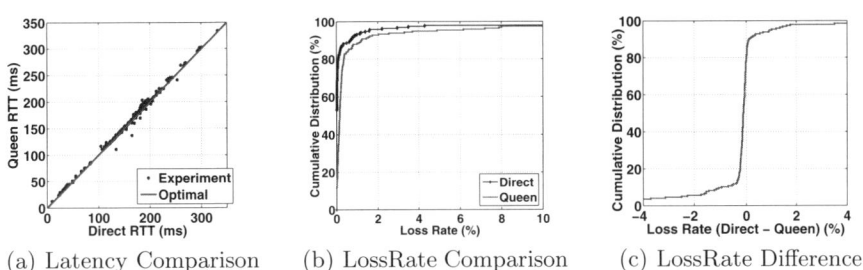

(a) Latency Comparison   (b) LossRate Comparison   (c) LossRate Difference

Fig. 6. PlanetLab Validation Results

within 1% for more than 80% of paths. Finally, it appears that iPlane does *not* return loss rate for any of the paths in this experiment. Apparently, these paths are not covered by its database.

4.2   PlanetLab Validation

In this set, we use PlanetLab (PL) to conduct validation, as in Fig. 4(b). Similar to 4.1, we again run two kinds of probes in parallel with the same pattern. The difference is: in direct probing, we run a UDP probing client at one PL node and a UDP echoing server at another PL node, to get the ground truth loss rate. We locate two nearby corresponding DNS servers (e.g. no more than 5ms away), one for each PL node, and estimate the loss rate between them. We compare loss rate estimated by both methods to see whether they match each other. In this set of validation, the path whose loss rate is estimated by Queen is slightly different from the path estimated by direct probing – the former path is between the two name servers while the later one is between the two PL nodes.

We choose 5 PL nodes with recursive local DNS servers as sources. Each source picks ~70 target PL nodes from different continents. We get results for ~260 paths, where 200 show loss either in UDP probing or Queen. Fig. 6 depicts the results. Again, two latencies and loss rates match very well. In particular, the absolute loss rate difference between two methods is within 1% for more than 85% paths. At the meantime, iPlane returns *zero* loss rates for all the paths in this experiment, which is quite far from both Queen and the ground truth.

Previous study [15] suggested that small 40-byte probes tend to experience less losses than large 1000-byte probes. Since query packets generated by Queen can be at most 280 bytes (limited by the maximum length of DNS hostnames, which is 255), it raises a concern that Queen might under-estimate the true loss rate. To study this issue, we re-run the same validation in parallel with different probing size of 80B, 160B and 240B. We observe that packet size in fact has very little effect on loss rate. In addition, we compare direct UDP probes with size of 80B, 240B and 1200B (very large, close to MTU) and the packet size also has very minimum effect(details skipped due to space constraint).

5   Experiment

After validating that our method has reasonably high accuracy, we conduct a loss rate measurement for a large geographic area with world-wide coverage.

5.1   Measurement Setup

We pick one server for each country from our open recursive DNS server list and measure the loss rate between each pair of servers. The final data set covers 6 continents and 147 countries, with 10,731 paths in total. The complete measurement involves a large number of paths. On each path, Queen sends query probes following exponential inter-arrival time with average 500ms for 15 minutes, so that their associated counting process is Poisson. To speed up the measurement process, we have developed a distributed execution platform, which splits the complete task into many smaller jobs, spreads these jobs onto PL nodes and executes them in parallel. This platform helps to complete the measurement quickly (e.g., with 300 PL nodes, 10,000 paths, each node only needs to run 33 jobs, and the entire task takes slightly more than 8 hours to complete).

5.2   Summary of Results

We group the sampled servers by continent and analyze loss rates within/cross continents. Fig. 7(a) shows the loss rate statistics within each continent, and Fig. 7(b) cross continents. Some results are intuitive – North America and Europe have low loss rates, no matter intra-continent or cross-continent. This is clearly due to good networking infrastructure with the two regions and connectivity between them. In addition, North America and Europe always have lower loss rates within the continent than cross to the other continent. Not as intuitive though, we also observe that, for other continents, the loss rates are in fact lower cross to North America or Europe than within the continent itself. This, we believe, reflects the fact that North America and Europe are currently the hubs of the Internet.

6   Related Work

The study of Internet packet loss rate can be dated back to more than a decade ago. It is conducted to understand Internet itself, as well as the impact on

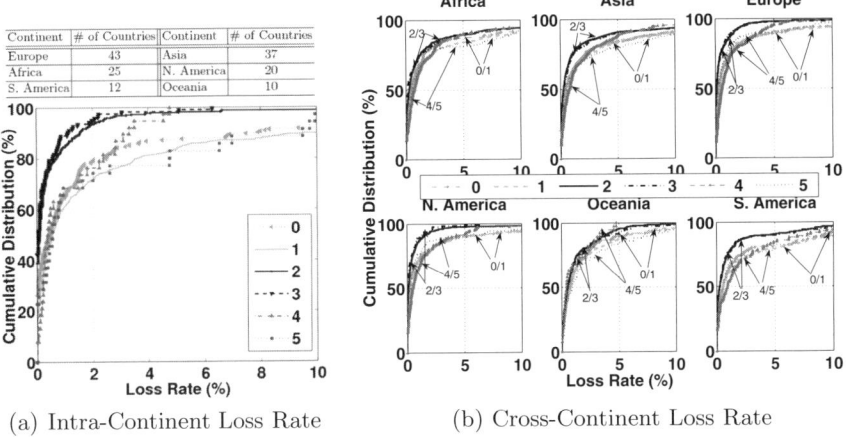

(a) Intra-Continent Loss Rate   (b) Cross-Continent Loss Rate

Fig. 7. Continent Loss Rate Statistics. Numbers 0-5 represent 6 continents, respectively. *0-Africa, 1-Asia, 2-Europe, 3-N.America, 4-Oceania, 5-S.America.*

the performance of applications [13,14]. Constant efforts are continuously being pushed to improve the accuracy of packet loss rate estimation [12]. Tools [15] are developed to use loss rate to troubleshoot path failures. All these work rely on sending out active UDP/ICMP probes. Hence, they require controlling of either one or both ends of the path being studied.

iPlane [1] is the only other tool *close* to be able to estimate packet loss rate without requiring access to either end. It constructs an annotated map of the Internet by: (*i*) sending probes from a large number of various vantage points, such as PlanetLab nodes and traceroute servers; (*ii*) clustering interfaces into PoPs based on response source address or returned TTLs to all the vantage points. By collecting all probes and processing the measurement data, it characterizes the loss rate of all inter-cluster links in the measured topology. Then, it may indirectly predict packet loss rate between a pair of end hosts by compounding the packet loss rate of each segment link along the path. However, iPlanes's coverage is limited as it can not provide packet loss rate on a path if neither end of the path exists in the database. Thus, it doesn't really provide loss rate between arbitrary two end-hosts, as it still requires contributions from one end. In addition, it does not perform measurement on demand.

7   Conclusion

In this paper, we presented *Queen*, a tool that estimates loss rate between arbitrary Internet end hosts without control of either side. We validate *Queen* with two different data sets. They all show that our method has reasonably high accuracy. We used *Queen* for an Internet-wide experiment, which provides informative results and realistic Internet characteristics.

References

1. Madhyastha, H.V., Isdal, T., Piatek, M., Dixon, C.: iPlane: An Information Plane for Distributed Services. In: USENIX OSDI (2006)
2. Gummadi, K.P., Saroiu, S., Gribble, S.D.: King: Estimating Latency between Arbitrary Internet End Hosts. In: ACM SIGCOMM IMW (2002)
3. EdgePlatform, Akamai Inc.,
   http://www.akamai.com/html/technology/index.html
4. Andersen, D.G., Balakrishnan, H., Kaashoek, M.F., Morris, R.: Resilient Overlay Networks. In: ACM SOSP (2001)
5. Yang, H.-Y., Lee, K.-H., Ko, S.-J.: Communication quality of voice over TCP used for firewall traversal. In: ICME (2008)
6. Huang, C., Wang, A., Li, J., Ross, K.W.: Understanding Hybrid CDN-P2P: Why Limelight Needs Its Own RedSwoosh. In: NOSSDAV (2008)
7. Ng, T.S.E., Zhang, H.: Predicting Internet Network Distance with Coordinates-Based Approaches. In: IEEE INFOCOM (2002)
8. Keynote Global Test and Measurement Network, Keynote Inc.,
   http://www.keynote.com/company/keynote_network/methodology.html
9. Dabek, R., Cox, R., Kaashoek, M.R., Morris, R.: Vivaldi: A Decentralized Network Coordinate System. In: ACM SIGCOMM (2004)
10. Ledlie, J., Gardner, P., Seltzer, M.: Network Coordinates in the Wild. In: USENIX NSDI (2007)
11. White, B., et al.: An Integrated Experimental Environment for Distributed Systems and Networks. In: USENIX OSDI (2002)
12. Sommers, J., Barford, P., Duffield, N., Ron, A.: Improving Accuracy in End-to-end Packet Loss Measurement. In: ACM SIGCOMM (2005)
13. Paxson, V.: End-to-end Routing Behavior in the Internet. In: ACM SIGCOMM (1997)
14. Bolot, J.C.: End-to-end Packet Delay and Loss Behavior in the Internet. In: ACM SIGCOMM (1993)
15. Mahajan, R., Spring, N., Wetherall, D., Anderson, T.: User-level Internet Path Diagnosis. In: ACM SOSP (2003)

Fast Available Bandwidth Sampling for ADSL Links: Rethinking the Estimation for Larger-Scale Measurements*

Daniele Croce, Taoufik En-Najjary, Guillaume Urvoy-Keller, and Ernst W. Biersack

EURECOM, Sophia Antipolis, France
{croce,ennajjar,urvoy,erbi}@eurecom.fr

Abstract. Most existing tools for measuring the end-to-end available bandwidth require access to both end-hosts of the measured path, which severely restricts their usability. Few tools have been developed to overcome this limitation, but all of them focus on achieving high precision and are not suitable for large campaigns. In this paper we develop *FAB-Probe*, a tool aimed at characterizing the available bandwidth of a large number of paths, adapting it particularly for ADSL settings. FAB-Probe is an evolution of *ABwProbe*, a tool that estimates the available bandwidth in non-cooperative ADSL environments. Analyzing carefully the needs of such a characterization tool, we optimize and rethink ABwProbe for larger-scale measurements. The validation of FAB-Probe is obtained both "in-lab", with ADSL hosts under our control, as well as under real traffic conditions, with the help of an ISP. Finally, as a proof of concept, we analyze the available bandwidth of over 1300 hosts participating to the KAD DHT used by eMule, periodically monitoring some static peers for over ten days.

1 Introduction and Motivation

Nowadays, over 221 million users access the Internet through broadband links and in many developed countries broadband technologies have reached more than half of the households, with peaks of over 90% penetration [10]. Moreover, subscribers are expected to double in the next four years, confirming the exponential increase observed until now [13]. Over 137 millions broadband users (62% of all subscribers) access the Internet through an ADSL link. Broadband technologies have allowed the development and the success of new bandwidth intensive applications (peer-to-peer applications for example). This have pushed the consumption of bandwidth and, therefore, the development of new, high capacity technologies. Despite these improvements, broadband technology is still the fundamental bottleneck for achieving higher performances [3].

Researchers had, up to now, limited access to information related to the characteristics of residential broadband networks. This was mainly caused by the difficulties in measuring these networks without explicit cooperation of the end hosts or the ISPs. Indeed, only few existing tools have been designed to work in non-cooperative[1] environments and *all focus on achieving high precision* while are not suitable for large

---

* This work is supported in part by the NANODATA-CENTERS program (FP7-ICT-223850) of the EU.

[1] Following the terminology used in [2, 11], with "non-cooperative" we mean that access to the remote host is not available.

S.B. Moon et al. (Eds.): PAM 2009, LNCS 5448, pp. 67–76, 2009.

measurement campaigns. Moreover, to the best of our knowledge, only ABwProbe [2] is tuned precisely for the available bandwidth (avail-bw) estimation of ADSL links.

In this paper, we propose FAB-Probe which is the natural evolution of ABwProbe. We rethink and optimize the tool for larger-scale measurements, giving particular attention to the *time needed for obtaining a good estimate*. Observing common traffic patterns, we elaborate a new strategy for sampling the avail-bw efficiently, adapting it to the absolute capacity of the ADSL. We reduce at minimum the number of probes and we refine existing cross-traffic filtering techniques. We also develop a method for detecting uplink congestion, which can alter the measurements. After validating FAB-Probe "in-lab" against ADSL hosts under our control, we test it in real traffic conditions with the collaboration of an ISP, obtaining very good results. Finally, as a proof of concept, we use FAB-Probe for measuring over 1300 peers participating in KAD, the DHT used by eMule: we measure the hosts closely in time so to obtain a snapshot of the avail-bw distribution, and we select 82 with static addresses that we measure periodically for over 10 days. We also provide FAB-Probe for public evaluation, available at [5].

2   Background and Related Work

2.1   Active Measurements in Non-cooperative Environments

In a cooperative environment, active measurements are usually done in the following way: a specific number of probing packets with appropriate characteristics (size, rate, etc.) are transmitted through the network, link or device, that has to be measured, towards a receiver. On the receiver side, the probes are captured and some metrics (latency, rate, inter-packet gap, etc.) are computed. By analyzing and comparing these metrics at the receiver, it is possible to infer some characteristics of the network traversed and to estimate the desired quantity.

In a non-cooperative environment, instead, there is no control on the receiving host and, thus, there is no way to analyze the probes received at the destination. Therefore, all statistical information on the probes must be obtained in a different and indirect manner. An idea that has been adopted in various tools [1, 2, 3, 6, 7, 11], is to send particular types of probes which should induce the receiving host to reply with some other packets, "echoing" the received probes. Examples of probes that have this quality are: ICMP Echo and Timestamp requests, most TCP packets (ACKs, SYNs, FINs, etc. but not RSTs), UDP packets sent to closed ports.

When undertaking measurements in this non-cooperative way, it becomes challenging to distinguish between the characteristics of the forward and of the reverse path: when measuring the forward path, we must verify that the observed metrics capture the characteristics of the forward path and not the characteristics of the reverse path (and viceversa). An important feature that makes TCP-based probes attractive, is that the packet sent back in reply by the non-cooperative host is a 40 byte packet (in most cases a RST) regardless of the size of the probe that has generated it. This property is fundamental for measuring the avail-bw: since packets traverse the network in both directions, the size of the probes must be adjusted to differentiate between forward and reverse path and interpret the results correctly. *In this work we focus on the estimation of the downlink avail-bw, as the uplink estimation is conceptually similar.*

2.2   Tools for Non-cooperative Available Bandwidth Estimation

While a multitude of utilities exist for cooperative environments, very few are designed to measure the avail-bw in settings where cooperation is not granted. Two tools, *pathneck* [7] and *Sprobe* [11], do not require access to the receiving host, however, the first one measures the Average Dispersion Rate (ADR, an upper bound of the avail-bw) and, since it is based on ICMP probes, can suffer from ICMP rate limiting; the second one is based on the Probe Gap Model which has been criticized and proven inaccurate in [8]. Some techniques dedicated to broadband networks (ADSL and Cable) are described in [3] for the estimation of capacity, buffer depth, queue management and others. However, the avail-bw is not taken into account.

To the best of our knowledge, the only tool designed especially to measure the avail-bw in non-cooperative environments is ABwProbe, which is particularly tuned to measure non-cooperative ADSL hosts. FAB-Probe is thus a natural evolution of ABwProbe which in turn is inspired by Pathload [4]. Both ABwProbe and FAB-Probe use the RTT in place of the One-Way Delay (OWD) of Pathload: if no cross-traffic disturbs the RSTs *on the reverse path*, indeed, an increasing trend in the OWDs will be preserved and will appear in the RTT samples as well. ACK probes are sent at rate R to the non-cooperative host which replies with fixed-size RSTs. The probes are only echoed by the receiver and at the sender the RTT is computed. This way, the RTT measured is used in place of the OWD (which is impossible to measure) for detecting if $R > A$ or not, A being the avail-bw of the path. However, since packets traverse the network a second time, the received RSTs will carry both information of the probes from the forward path, but will also be influenced by the characteristics of the reverse path. This becomes an important problem in asymmetric environments such ADSL because RSTs have to traverse the uplink of the ADSL which has much lower capacity (and usually lower avail-bw) than the downlink.

To overcome this asymmetry, in ABwProbe (and in FAB-Probe as well) we exploit the fact that TCP RSTs are always 40 bytes long regardless of the size of the ACKs: if the probes are very large, say 1500 bytes, then the rate generated on the downlink will be $1500/40 = 37.5$ times higher than the load on the uplink (Layer-2 overheads excluded, see [2]). Now, most ADSLs have uplink capacities that are less than 8 times lower than the downlink [3] so this measure should be sufficient to overcome the *capacity* asymmetry. In FAB-Probe an additional test is implemented to detect when the uplink *avail-bw* is insufficient for the RSTs to be transmitted on the uplink.

Finally, even when the avail-bw on the uplink is sufficient, it has been shown in [2] that large cross-traffic packets interfering on the uplink can significantly affect the measurement because of the compression of the RSTs queued behind. Since an MTU packet can take several tens of milliseconds to be transmitted, this effect is not negligible. In ABwProbe some techniques were proposed to detect and filter RTT samples affected by RST compression. As we will explain later, in FAB-Probe we improve the filtering by combining the proposed methods to better exploit their strengths.

Like other tools, ABwProbe too is aimed at providing very high precision in estimating the avail-bw. In FAB-Probe we intentionally renounce to some measurement resolution, in favor of the measurement latency which is remarkably reduced.

3   FAB-Probe: Fast Available Bandwidth Sampling for ADSL Links

3.1   What Should We Measure Exactly?

Most tools measuring the avail-bw aim at providing an *absolute* estimation regardless of the capacity of the link. This makes sense for applications (such as video stream-ing) that require a certain amount of bandwidth to work properly. In general, however, to characterize a network and the load generated by the users, it is more interesting to know the avail-bw *relative* to the absolute capacity. For example, it is often more useful to know that the bottleneck link has 30% left of its spare capacity rather than knowing there are 200 kbps available, as the first metric intrinsically provides an idea of link quality while the latter does not. Additionally, for most large-scale characteriza-tion purposes, it is critical to have a fast estimation process while it is less important to have extremely high precision. For example, knowing that the avail-bw is in between 90% and 100% of the total capacity is usually enough for classifying the link as *inac-tive*. This aspect is very important because most estimation tools (including ABwProbe) reiterate the probing phase until the desired precision is obtained, i.e. higher precision translates in longer running time. The intuition behind FAB-Probe is thus to analyze the avail-bw relative to the capacity, probing the avail-bw at few key values (only 5 by default) and providing a quick, but still accurate, estimate of the avail-bw of the link. In this paper the focus is on the avail-bw estimation and discussions on how to estimate the capacity are out of its scope. For a description of the capacity estimation technique used in FAB-Probe the interested reader can refer to [1].

Focusing on larger-scale measurements, it is important to note that the large major-ity of users are either *inactive*, i.e. not using their bandwidth, or *active*, meaning they are actively using their connection. This active-inactive behavior has already been ob-served in [14] and is confirmed by our experimental results. Recalling that the network bottleneck has been shown to be at the edge of the network [3], there are high chances that active periods tend to bring the ADSL close to congestion. We can thus design the avail-bw estimation algorithm exploiting this *a priori* information. In particular, with an active-inactive pattern, the measurements should be more fine-grained on extreme avail-bw values (closer to 0 and 100%) while can be less precise on "middle" values. In FAB-Probe we thus limit the exploration of the avail-bw range by testing only few key values, distributed in the following way: 10%, 25%, 50%, 75% and 90% and the avail-bw will be captured in between two of these values (and 0 and 100% but these values are obviously not probed). This guarantees that the number of probing itera-tions is always less than 5 (see details below), achieving a considerable speedup of the measurement process while limiting the probing to significant values of the avail-bw range. Clearly, the precision can be increased (or reduced) by changing the number of predefined values and also the distribution of the values itself can be changed to better fit the avail-bw distribution – if this is available. Finally, note that the *absolute* avail-bw can still be obtained from the relative values of FAB-Probe because the capacity is known.

3.2 Measurement Algorithm and Speedup

In FAB-Probe we first measure the downlink capacity C with the techniques in [1]. Then, we explore the avail-bw range with a binary search-like algorithm, modifying the one in ABwProbe: the path is probed with a fleet of packets at an initial rate $R = 75\% \times C$ and the RTT of the packets is analyzed to detect if there is an *increasing trend*. An increasing trend would indicate self-induced congestion on the path, thus meaning that $R > A$. The rate is then reduced (increased) if an increasing trend (no trend) in the RTTs is detected. Consequently, R is updated to a lower (higher) value, i.e. 50 or 90%, and another fleet is sent. The process is iterated until the avail-bw is captured between two probed values, where one has shown an increasing trend, the other one no trend. For example, if $A = 60\% \times C$, the first fleet at 75% would have an increasing trend, while the second fleet at 50% would show no trend. Since there are no other measuring values in between 50 and 75%, the algorithm stops and FAB-Probe would output an avail-bw range of 50-75%. The main difference with previous algorithms here, is that instead of changing R in a pure binary search fashion, in FAB-Probe the probing rate R assumes few deterministic values, significantly reducing the running time.

To further speedup the measurement time, we have worked also on the number of fleets to send. In Pathload, to detect if there is an increasing trend or not, 12 independent fleets are sent at the same rate R. Then, the trend is computed on all these fleets and the decision if $R > A$ or not is taken based on the fraction of fleets agreeing. This process is repeated for all probing rates regardless of the trend intensity. In [2] we already proposed a strategy to accelerate the decision process, reducing on-the-fly the number of fleets when the trends are particularly pronounced, thus adapting the number of fleets to the difference between R and A. In FAB-Probe we push the idea to the limit as we simply do not investigate bandwidth rates that are too close to the avail-bw. The decision is taken based on the result of one fleet only: if the trend is clearly noticeable, the measurement is valid and the rate R is reduced (increased) right away. Otherwise, if the trend is not pronounced, this means that the avail-bw is close to the rate R, the fleet is marked as "grey", and two other fleets are sent at higher and lower rate just as a counter check. The rationale behind this is the following: suppose the avail-bw is close to 70% of the capacity. If the probing rate R is at 75% of the capacity, there will be a very weak increasing trend in the RTT samples. In a situation like this, ABwProbe would send more and more fleets (up to 12) to have good confidence before declaring that $R > A$, getting "stuck" on this rate value. In FAB-Probe instead, we immediately probe the link at 50% and 90% (the values just above and below 75%) and, if the avail-bw is confirmed to be in between, the range given in output would be 50-90% – instead of 50-75%. This way we obtain an acceptable result with very few fleets (only 3 in the example), trading off precision for running time, and obtaining a great speedup.

3.3 Uplink Congestion and Cross-Traffic

Suppose an increasing trend in the RTTs is detected while measuring the downlink avail-bw. Since the estimation is done in a non-cooperative environment, we must verify that the observed increase in RTT is due to the self-induced congestion on the downlink and not to compression of the RSTs on the reverse path. The uplink of the ADSL, in particular, is critical because of the very low bandwidth that can be easily saturated.

For this reason, after the first increasing fleet is detected, we make this simple test to check the conditions of the uplink: we send a second fleet in which the probes have the same spacing between each other but the size is at minimum (40 Bytes). This fleet reproduces on the reverse path a sequence of RSTs with the same *rate* of the fleet before (same size, same gap), however, on the downlink the rate is at least 16 times lower (more if ATM is not adopted, see [2]). Now, if this fleet still shows an increasing trend either (i) the downlink is completely congested, or (ii) the uplink is loaded and the RSTs on the reverse path are compressed. In both cases, however, estimating the avail-bw precisely in these critical conditions becomes problematic (also in terms of measuring time because the rates are extremely low) therefore we classify the ADSL as "congested" without investigating further.

In FAB-Probe we also refine the filtering methods presented in [2]. Indeed, uplink cross-traffic can compress the RSTs on the uplink, even without serious congestion, generating spikes in the RTT that temporarily "hide" the temporal properties of the downlink. In [2] two techniques are proposed, one statistical based on a robust method called Iteratively Re-wighted Least Squares, and one deterministic, aimed at detecting the consecutive decreasing RTT samples resulting from the compression of the RSTs. The two methods have interesting properties but also downsides: the statistical method suffers in high cross-traffic conditions while the deterministic one is ineffective when the cross-traffic packets are small. In FAB-Probe we combine the two methods to benefit from both approaches: first we filter the large majority of affected RTT samples detecting consecutive decreasing trends, then we filter the remaining samples applying the robust statistical method. As we will show, this allows us to obtain correct measurements even when the uplink is highly loaded.

4   Validation

To validate the accuracy of FAB-Probe, we have tested our tool both on some ADSL hosts under our control and against hosts belonging to an ADSL service provider from which we obtained the traffic traces as "ground truth". The measuring host was always a well-connected host, with a 100Mbps connection and with at least 90 Mbps avail-bw, so that the bottleneck link was the ADSL on the other edge of the network.

4.1   In-lab Validation

We tested the accuracy of FAB-Probe in different traffic conditions and with various cross-traffic rates. From a third well-connected host, we used a traffic generator to inject CBR traffic towards the ADSL host. We vary the load on the downlink spacing several rates close to the ones used by FAB-Probe to probe, so to verify that no error occurs when R and A are close – the worst case. For example, we test FAB-Probe with 45% and 55% avail-bw because FAB-Probe runs fleets at 50%, thus verifying if the trend is correctly dectected. Tab. 1 summarizes the results given in output by FAB-Probe as well as the total number of fleets sent and the running time. The improvement compared to ABwProbe is remarkable: the number of fleets is an order of magnitude lower, thus reducing significantly the intrusiveness of the measurement, and the avail-bw estimate is obtained over 5 times faster while still providing precise avail-bw ranges.

Table 1. Accuracy and speed of FAB-Probe: the output range effectively captures the avail-bw. Compared to ABwProbe (ABP), the number of probing fleets and the running time are considerably lower. The asterisk indicates that an additional fleet with 40 bytes probe packets was sent to verify uplink congestion.

		Downlink avail-bw range (relative to capacity)							
	Real avail-bw	95%	85%	70%	55%	45%	30%	20%	5%
FAB-P.	Measured range	90-100%	75-90%	50-75%	50-75%	25-50%	25-50%	10-25%	cong.
	# of fleets	2	2*	3*	3*	3*	3*	3*	1*
	Time (sec)	5.6	7.6	14.4	14.5	13.5	14.7	23.0	6.0
ABP	# of fleets	17	19	21	24	23	18	21	16
	Time (sec)	31.8	40.2	57.4	68.4	70.3	56.9	88.3	200.9

Table 2. Impact of uplink cross-traffic: FAB-Probe correctly measures the avail-bw in presence of cross-traffic both when the downlink is idle and loaded (30% avail-bw)

	Uplink cross-traffic (load relative to uplink capacity)							
	0%	15%	25%	35%	45%	55%	65%	75%
Measured range (100% avail-bw)	**90-100%**	90-100%	90-100%	90-100%	75-100%	75-100%	75-90%	cong.
Measured range (30% avail-bw)	**25-50%**	25-50%	25-50%	25-50%	25-50%	25-50%	10-25%	cong.

To prove the robustness of FAB-Probe and the effectiveness of the uplink cross-traffic filtering, we have measured the downlink both in idle and loaded conditions while injecting increasing cross-traffic on the uplink. We generated CBR traffic from the ADSL host towards a third machine using MTU size packets (thus causing the highest RST compression, see [2]). As shown in Tab. 2, the cross-traffic filtering allows FAB-Probe to correctly estimate the downlink avail-bw both when the link is idle and loaded at 70%, and the congestion test operates when the cross-traffic is too high (over 450 kbps in the case shown, with an uplink of 578 kbps). Only a slight underestimation occurs in extreme cross-traffic conditions.

4.2 Validation "In the Wild"

To further test the accuracy of FAB-Probe, we validated the tool in real-world traffic conditions. With the help of an ADSL service provider, we first selected few hundred hosts which were not idle nor completely saturated, in order to have some avail-bw variability. Then, we periodically measured 13 ADSL hosts during a one hour period. The ISP provided us the time series of the incoming and outgoing traffic of the measured hosts and we could thus compute the amount of traffic actually traversing the ADSL. The results obtained matched very well the ones provided by FAB-Probe and two interesting examples are provided in Fig. 1. In Fig. 1(a), FAB-Probe was precisely measuring the avail-bw, being conservative and widening the range given in output when the probing rate was too close to the real value ("grey" trend). In Fig. 1(b) the host measured is a bit more loaded and the variability of the avail-bw process is higher. Nevertheless, FAB-Probe provides very good estimates, with only one inaccurate sample.

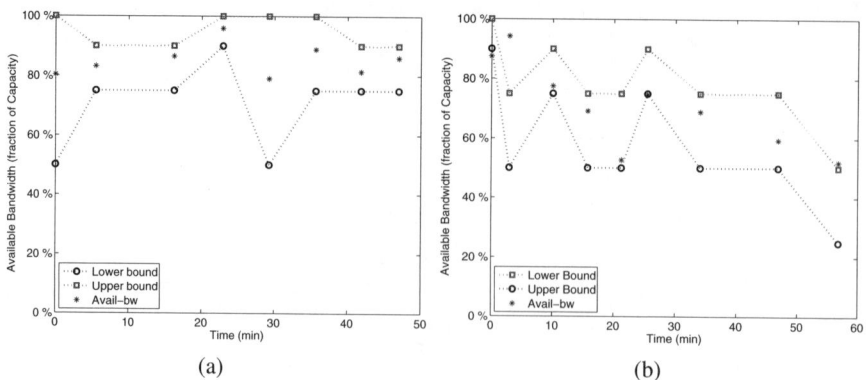

(a)   (b)

Fig. 1. Real-world traffic conditions: FAB-Probe effectively captures the avail-bw of the hosts

5   Experimental Results on KAD

As a proof of concept, in this section we show the results obtained with FAB-Probe on over 1300 ADSL hosts participating to the Kademlia DHT [9] used by applications like eMule. We used the KAD crawler of [12] to retrieve the IP addresses of the KAD hosts and then selected with the Maxmind database only those connecting through an ADSL service provider. We have run two different experiments: one aiming at analyzing a large variety of hosts closely in time, generating a snapshot of the avail-bw distribution, and another measuring a particular subset of hosts for over 10 days.

5.1   A Snapshot of the Avail-bw Distribution

We have measured 1244 hosts from various ISPs in US and Europe analyzing both capacity and avail-bw (although we omit the discussion on the capacity, see [1] on this matter). Avail-bw measurements lasted on average less than 6 seconds per host. Fig. 2(a) shows the downlink avail-bw range for these hosts. Confirming the results in [14], hosts are sharply partitioned in two groups, *idle* or *active*, since more than 80% have either

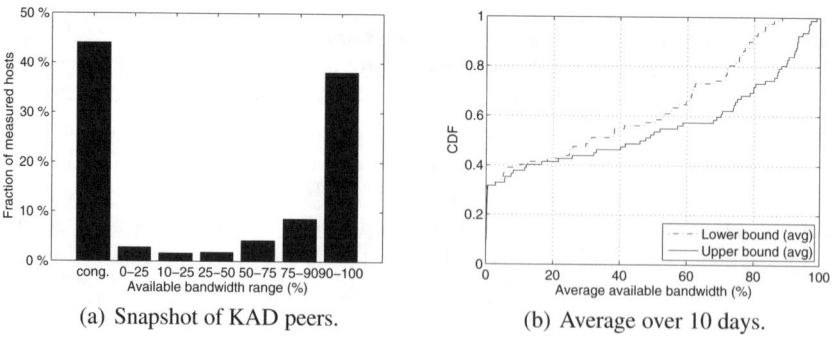

(a) Snapshot of KAD peers.   (b) Average over 10 days.

Fig. 2. Avail-bw in KAD: snapshot (a) and 10 days average (b) of the downlink avail-bw

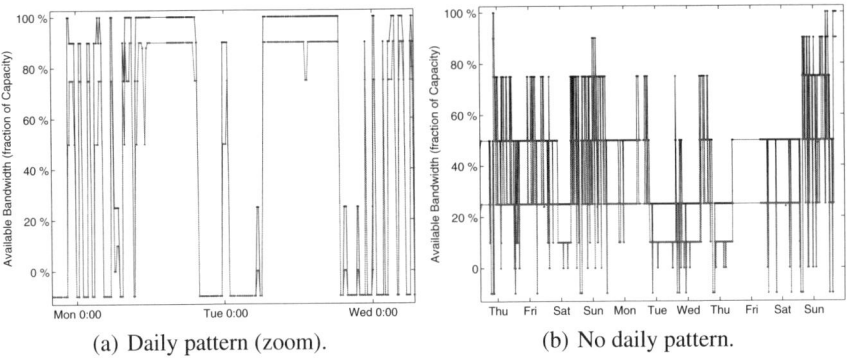

(a) Daily pattern (zoom).   (b) No daily pattern.

Fig. 3. Avail-bw evolution over time: comparison of two hosts that had different avail-bw patterns. Negative values indicate that the uplink was congested.

over 90% avail-bw or a congested uplink[2]. Less than 2% of the hosts measured had a fleet with unclear trend ("grey" in Pathload) resulting in a wider avail-bw range, such as 50%-90%, so we omitted them from the figure.

5.2 A Ten-day Case Study

From the hosts participating in KAD, we selected the ones that were continuously online for over one month using the same (static) IP address. These hosts are a good opportunity for a long-term analysis of the avail-bw evolution. However, they constitute a particular subset of eMule hosts that might continuously be seeding the file distributions acting as traditional servers. We have selected 82 of these hosts and measured them every 5 minutes for over 10 days. The capacity was also monitored to make sure it did not change during that period. In Fig. 2(b) are shown the *average* avail-bw lower and upper bounds considering the entire measurement period. Over 30% of the hosts had the uplink continuously congested meaning they were always uploading large amounts of data. Over 25% of the hosts were instead frequently idle, as on average, they had *at least* 75% avail-bw. We emphasize that the distribution is less sharp than the snapshot in Fig. 2(a) because Fig. 2(b) shows the CDF of averages computed over almost 3000 estimates per host, while the latter is a one shot measurement.

Analyzing more closely the results, some hosts showed sharp daily patterns in the avail-bw, while others did not. For example, Fig. 3(a) shows a host that was basically idle during night hours while was active during the day, while Fig. 3(b) shows another host which showed more slow-varying patterns during the measurement campaign. In general, however, this set of hosts showed little change in their avail-bw between day and night, yet maintaining an active-inactive pattern with sometimes steep variations.

---

[2] Modern applications for P2P file distribution employ "tit-for-tat" algorithms that encourage the the peers to actively participate to the distribution process. This means that peers upload data while downloading, thus using their bandwidth in a more symmetrical fashion compared to client-server applications. Now, the uplink of an ADSL being much smaller than the downlink, this explains why eMule peers suffer from uplink congestion before the downlink is saturated.

6   Conclusions

In this paper we presented FAB-Probe, a new measurement tool for non-cooperative ADSL environments, designed for larger-scale measurements. Rethinking previous techniques and studying the requirements in terms of accuracy and time, FAB-Probe is the evolution of ABwProbe, from which the basic algorithm was acquired. We developed a new sampling strategy, balancing accuracy and measuring time to sample efficiently the avail-bw; we minimized the number of probing fleets; we studied a method to detect uplink congestion; and we improved the cross-traffic filtering technique. We proved the accuracy of FAB-Probe both "in-lab", against ADSL hosts under our control, and in real traffic conditions, with the cooperation of an ISP. Finally, as a proof of concept, we measured with FAB-Probe over 1300 ADSL hosts participating in KAD, for a snapshot of the avail-bw distribution as well as a characterization of 82 hosts for over ten days.

Acknowledgments

We are extremely thankful to Marcin Pietrzyk for providing us the traces used in the validation and to Moritz Steiner for his help in obtaining the IPs of the KAD users.

References

1. Croce, D., En-Najjary, T., Urvoy-Keller, G., Biersack, E.: Capacity Estimation of ADSL links. In: CoNEXT (December 2008)
2. Croce, D., En-Najjary, T., Urvoy-Keller, G., Biersack, E.: Non-cooperative Available Bandwidth Estimation towards ADSL links. In: Proc. Global Internet Symposium 2008 (April 2008)
3. Dischinger, M., Haeberlen, A., Gummadi, K.P., Saroiu, S.: Characterizing residential broadband networks. In: IMC (October 2007)
4. Dovrolis, C., Jain, M.: End-to-end available bandwidth: Measurement methodology, dynamics, and relation with TCP throughput. In: SIGCOMM, Pittsburgh, USA (August 2002)
5. The FAB-Probe Project website,
   http://www.eurecom.fr/~btroup/fabprobe.html
6. Haeberlen, A., Dischinger, M., Gummadi, K.P., Saroiu, S.: Monarch: A tool to emulate transport protocol flows over the internet at large. In: IMC (October 2006)
7. Hu, N., Li, L.E., Mao, Z.M., Steenkiste, P., Wang, J.: Locating internet bottlenecks: algorithms, measurements, and implications. In: SIGCOMM, Portland, USA (2004)
8. Lao, L., Dovrolis, C., Sanadidi, M.Y.: The probe gap model can underestimate the available bandwidth of multihop paths. Computer Communication Review 36(5), 29–34 (2006)
9. Maymounkov, P., Mazieres, D.: Kademlia: A peer-to-peer information system based on the XOR metric. In: Druschel, P., Kaashoek, M.F., Rowstron, A. (eds.) IPTPS 2002. LNCS, vol. 2429, p. 53. Springer, Heidelberg (2002)
10. OECD. OECD broadband statistics (June 2007),
    http://www.oecd.org/sti/ict/broadband
11. Saroiu, S., Gummadi, P.K., Gribble, S.D.: Sprobe: A fast technique for measuring bottleneck bandwidth in uncooperative environments (2002),
    http://sprobe.cs.washinton.edu
12. Steiner, M., En-Najjary, T., Biersack, E.W.: A Global View of KAD. In: IMC (2007)
13. Windsor, L.: Oaks Group. Annual market outlook report (2006)
14. Zaragoza, D., Belo, C.: Experimental validation of the on-off packet-level model for ip traffic. Comput. Commun. 30(5), 975–989 (2007)

Multi-layer Monitoring of Overlay Networks[*]

Mehmet Demirci[1], Samantha Lo[1], Srini Seetharaman[2], and Mostafa Ammar[1]

[1] School of Computer Science, Georgia Institute of Technology, Atlanta GA 30332
[2] Deutsche Telekom R&D Lab, Los Altos CA 94022, USA
{mdemirci,samantha,ammar}@cc.gatech.edu, srini.seetharaman@telekom.com

Abstract. Monitoring end-to-end paths in an overlay network is essential for evaluating end-system performance and for troubleshooting anomalous behavior. However, conducting measurements between all pairs of overlay nodes can be cumbersome and expensive, especially in a large network. In this paper, we take a different approach and explore an additional degree of freedom, namely, monitoring native links. We allow native link measurements, as well as end-to-end overlay measurements, in order to minimize the total cost of monitoring the network. We formulate an optimization problem that, when solved, identifies the optimal set of native and overlay links to monitor, and a feasible sequence of arithmetic operations to perform for inferring characteristics of the overlay links that are not monitored directly. We use simulations to investigate how various topological properties may affect the best monitoring strategy. We also conduct measurements over the PlanetLab network to quantify the accuracy of different monitoring strategies.

1   Introduction

Monitoring all links in infrastructure overlay networks with persistent nodes is necessary to assess the overall performance of the users and to detect anomalies. Since an *overlay link* is in reality an end-to-end native path spanning one or more native links, this full monitoring operation can constitute a significant overhead (in terms of bandwidth and processing) for large overlays, especially if the monitoring is performed by active measurements.

In this paper, we alleviate the overlay network monitoring problem by adopting a more flexible approach that allows certain native link measurements in addition to end-to-end measurements[1]. These native link measurements can be used to infer desired metrics for overlay links by suitable combinations of native layer metrics. We call this approach *multi-layer monitoring*. This framework allows for four different options:

1. **Monitor all overlay links:** With this strategy, all overlay links are monitored directly and individually.

---

[*] This work was supported in part by NSF grant CNS-0721559.
[1] Our work pertains to infrastructure overlays, rather than peer-to-peer networks.

S.B. Moon et al. (Eds.): PAM 2009, LNCS 5448, pp. 77–86, 2009.

2. **Monitor a *basis set* of overlay links:** The work in [8] introduces a method to select and monitor a minimal subset of overlay links called the *basis set*. The characteristics of the remaining overlay links are inferred from the measurements for the basis set.
3. **Monitor all native links:** Another option is to monitor all the underlying native links in the network. Afterwards observed native layer metrics are combined to produce the results for all the overlay links.
4. **Monitor a mix of native links and overlay links (Multi-layer Monitoring):** In this option proposed in this paper, we monitor some native links and a subset of the overlay links. We then infer the remaining overlay links by combining these observations.

Note that while options 2-4 have the potential to reduce the monitoring cost, they are also prone to *inference errors* when an overlay link measurement is inferred from measurements on native and/or other overlay links.

The multi-layer monitoring strategy (option 4) is the most general one and subsumes all others. It also affords significant flexibility in monitoring overlays. Our objective in this work is to minimize monitoring cost by determining the optimal mix between overlay and native layer monitoring. To this end we formulate this as an optimization problem and discuss some features of its solution.

Previous work has considered overlay network monitoring and developed various approaches for it. Chen et al. [8] propose an algebraic approach to efficiently monitor the end-to-end loss rates in an overlay network. They use linear algebraic techniques to find a minimal *basis set* of overlay links to monitor and then infer the loss rates of the remaining ones. iPlane [4] predicts end-to-end path performance from the measured performance of segments that compose the path. We generalize these techniques and allow measuring both end-to-end paths and underlying segments. Our approach in this paper requires a deep collaboration between the overlay network operator and the native network, similar to the design goals of the overlay-friendly native network[7].

The remainder of this paper is organized as follows: We describe the multi-layer monitoring problem in Section 2. Section 3 presents our linear program based solution. We present details from simulating the multi-layer monitoring framework in general topologies in Section 4. Section 5 describes PlanetLab experiments that we conducted to characterize the inference errors that can result from this multi-layer monitoring solution. We conclude the paper in Section 6.

2   The Multi-layer Monitoring Problem

We model the native network as a directed graph $G = (V, E)$, where V is the set of vertices and E is the set of directed edges connecting these vertices. Next, we model the overlay network as a directed graph $G' = (V', E')$, with $V' \subseteq V$ being the set of overlay nodes and E' being the set of overlay links. In a multi-layer network, each overlay link spans one or more native links. Thus, the following relation holds: $e' \in E'$ is a set $\{e_{1e'}, e_{2e'}, ..., e_{ne'}\}$, where $e_i \in E$ and $e_{ke'}$ denotes the kth native edge in e'.

Link monitoring incurs a certain cost, typically in the form of resource overhead (e.g., processor utilization, bandwidth), at each layer. We use $C(e)$ and $C'(e')$ as the cost of monitoring a native link and an overlay link respectively. Since $C(e)$ and $C'(e')$ are variables, the cost structure is flexible and can accommodate various scenarios. For instance, if it is not possible to monitor certain native links directly, the cost variables for those links can be set to infinity.

Let $\mathcal{M} = \{\mathcal{M}_1, \mathcal{M}_2, \ldots, \mathcal{M}_N\}$ represent the desired set of monitoring operations we would like to get results for, which in our case is the set of desired overlay link measurements. Let $\mathcal{P} = \{\mathcal{P}_1, \mathcal{P}_2, \ldots, \mathcal{P}_R\}$ represent the set of monitoring operations that are actually performed. This set can contain a mixture of native and overlay link measurements. Let *composition rule* $\mathcal{F}(\mathcal{P}, \mathcal{M}_i)$ represent a function that combines the results from available native and overlay link measurements to infer the desired measurement of the overlay link $\mathcal{M}_i$. In this work, we use the composition rule of the latency metric.

We say that a certain $\mathcal{M}$ is *feasible with respect to* $\mathcal{P}$, if all values in $\mathcal{M}$ can be computed from $\mathcal{P}$. Clearly, if $\mathcal{M} \subseteq \mathcal{P}$, then the monitoring problem is *feasible*. In cases when $\mathcal{M} \nsubseteq \mathcal{P}$, feasibility is not always assured.

The optimization problem can thus be stated as, "Given a monitoring objective $\mathcal{M}$, find the $\mathcal{P}$ such that $\mathcal{M}$ is feasible with respect to $\mathcal{P}$ and $cost(\mathcal{P}) = \sum_{i=1}^{R} cost(\mathcal{P}_i)$ is minimal."

Assumptions and Limitations. In this paper, we assume that the best-effort routing at the native layer treats measurement probes in the same manner as other data packets, so as to obtain an accurate estimate of the user experience. We restrict our work to the metric of latency, although it has been shown that the logarithm of link loss rates are additive metrics that can be composed in a manner similar to link latencies[8]. Furthermore, the linear programming formulation in

Table 1. Notations used

E	Edges in the native layer
E'	Edges in the overlay layer
$C(e)$	Cost to monitor native link e
$C'(e')$	Cost to monitor overlay link e'
$X_m(e)$	1 if native link e is monitored, 0 otherwise*
$X_i(e)$	1 if native link e is inferred, 0 otherwise*
$Y_m(e')$	1 if overlay link e' is monitored, 0 otherwise**
$Y_i(e')$	1 if overlay link e' is inferred, 0 otherwise**
$f(e, e')$	1 if overlay link e' is routed over native link e, 0 otherwise
$x_i(e, e')$	1 if native link e is inferred from overlay link e', 0 otherwise
$l_i(e)$	Integer representing the inference dependency between native links to resolve inference loops

* A native link can be monitored or inferred but never both. Some are neither monitored nor inferred if they are not needed in inferring overlay link measurements.

** An overlay link is either monitored or inferred, but never both.

the subsequent section cannot be applied for multi-path routing at the native layer: The overlay link latency composition rule needs revision for handling multi-path routing. We reserve these extensions to the model for future study.

3   Linear Programming Formulation

Using the notation presented in Table 1, we formulate the optimization problem as the following Integer Linear Program (ILP):

$$\textbf{minimize Total Cost} = \sum_{e \in E} X_m(e) \cdot C(e) + \sum_{e' \in E'} Y_m(e') \cdot C'(e') \quad (1)$$

subject to the following constraints

$$\forall\, e' \in E', e \in e' : \quad X_m(e) + X_i(e) = 1, \text{ if } (Y_m(e') + Y_i(e')) = 0 \;. \quad (2)$$

$$\forall\, e' \in E', e \in e', d \in (e' - e) : \quad x_i(e, e') \le (X_m(d) + X_i(d)) \;. \quad (3)$$

$$\forall\, e' \in E' : \quad \sum_{e \in e'} x_i(e, e') \le (Y_m(e') + Y_i(e')) \;. \quad (4)$$

$$\forall\, e \in E : \quad X_i(e) \le \sum_{e' \in E'} x_i(e, e') \le 1 \;. \quad (5)$$

$$\forall\, e' \in E', e \in e', d \in (e' - e) : \quad x_i(e, e') = \begin{cases} 1, & \text{if } l_i(e) > l_i(d), \\ 0, & \text{otherwise} \;. \end{cases} \quad (6)$$

$$\forall\, e' \in E' : \quad Y_i(e') = 1, \text{if } e' \text{ can be inferred from other overlay links in } \mathcal{P} \;. \quad (7)$$

$$\forall\, e \in E, e' \in E' : \quad X_m(e) \in \{0, 1\}, X_i(e) \in \{0, 1\}, x_i(e, e') \in \{0, 1\}, \\ Y_m(e) \in \{0, 1\}, Y_i(e) \in \{0, 1\} \;. \quad (8)$$

Constraints (2) to (8) assure the feasibility of the solution. These constraints can be explained as follows:

(2) This constraint, applied to all overlay links, determines the exact layer at which each overlay link is to be monitored. If the overlay link is not already monitored or inferred, then monitor, or infer, all native links it spans. Furthermore, this constraint will ensure that we only monitor or infer, and never both. This condition also prevents an overlay link from being monitored, if all its constituent native link measurements are already known.

(3) We enforce the constraint that a native link e is inferred from an overlay link e' only if all other native links in that overlay link are already monitored or inferred. This insures that the inferred native link can be appropriately calculated from other link measurements.

(4) This constraint insures that a native link e is inferred from an overlay link e' only if the overlay link latency is already monitored, or inferred, at the overlay layer (i.e., $Y_m(e') + Y_i(e') = 1$). Furthermore, we place the constraint that no more than 1 native link can be inferred from each overlay link. This is typically achieved in an ILP by setting the sum of individual variables $x_i(e, e')$ to be less than or equal to 1.

(5) This is a complex constraint which achieves three sub-goals: (a) Mark a native link as *inferred* if it is inferred on any of the overlay links that span it, (b) Mark a native link as *not inferred* if it is not inferred on any of the overlay links that span it, and (c) Insure that a native link is inferred only from 1 overlay link, so as to reduce wasting resources on performing multiple inferences. These three constraints ensure that we accurately mark a native link as *inferred*.

(6) This constraint is crucial to remove any *circular inference*, which can happen if we infer one native link measurement through an arithmetic operation on the measurement of another. We achieve this by assigning integer inference levels (denoted by variable l_i), such that a native link must be inferred only from other native links that have a lower inference level.

(7) We use this constraint to implement the basis set computation and infer some overlay link measurements from other known overlay link measurements.

(8) Lastly, we specify the binary constraints for all variables used. This constraint makes the problem hard.

We apply the above ILP to any given topology and solve it using the GNU linear programming kit[3], which uses the branch-and-bound approximation technique. The optimal solution for a given topology identifies the overlay links that can be inferred from other native and overlay links, and describes how these inferences should be done. Using this information, we infer the latency of all overlay links ($\mathcal{M}$) from available measurements ($\mathcal{P}$) in our database.

4 Examples Using Multi-layer Monitoring

In this section, we present various simulation experiments to demonstrate the types of results obtainable from our optimization approach and how it is affected by various network features. Although we only simulate intra-domain topologies, our model and ILP are equally applicable to multi-domain topologies.

Random Placement. In the first experiment we consider five native link topologies derived from Rocketfuel [6] data. For each network we generate an overlay network using approximately 20% the number of nodes in the native topology as overlay nodes. These nodes are placed randomly among the native nodes and fully-connected to form the overlay network. In this case, we define the cost of monitoring as the total number of native and overlay measurements needed. We consider the following four monitoring strategies:

– *Monitoring all overlay links:* The total cost is the cost of monitoring all $N.(N-1)$ overlay links, where N is the number of overlay nodes.

- *Monitoring all native links:* The total cost is the number of distinct native links spanned by all the overlay links.
- *Monitoring a basis set of overlay links:* To obtain this solution, we set the cost of monitoring a native link very high in our ILP so that the solution selects only overlay links for monitoring.
- *Monitoring a combination of native and overlay links:* We set the cost of monitoring a native link equal to the cost of monitoring an overlay link in the ILP. (From here on, we refer to these costs as *unitNativeCost* and *unitOverlayCost*, respectively.) The ILP then produces a solution that minimizes the total cost, which is the same as minimizing the number of measurements in this case.

Table 2 demonstrates the lowest total monitoring cost that can be achieved by the above monitoring strategies for each topology. In addition, the cost that results from monitoring native links and the cost that results from monitoring overlay links are reported separately for the multi-layer combination strategy in the last column. In all topologies, monitoring a combination of native and overlay links provides the lowest-cost option. On average, this lowest cost is 71% lower than the cost for the naive all-overlay approach and 11% lower than the all-native solution. This represents significant saving, while being flexible enough to accommodate other constraints.

Table 2. The lowest cost for each strategy when *unitNativeCost* = *unitOverlayCost*

AS #	Number of overlay nodes	All overlay	All native	Basis set	Combination (n: native, o: overlay)
1221	21	420	102	198	98 (66 n, 32 o)
1755	17	272	112	98	92 (42 n, 50 o)
3257	32	992	240	500	222 (142 n, 80 o)
3967	15	210	98	138	78 (46 n, 32 o)
6461	28	756	224	394	210 (146 n, 64 o)

Amount of link-level overlap. In this section, we study the effect of overlap between overlay links over the optimal monitoring solution. As a measure, we use the average number of overlay links that span a native link in the network. We call this value the *overlap coefficient*. For this analysis we use the results from the first experiment.

Table 3 demonstrates how the lowest cost solution, as given by our ILP, varies with the amount of link-level overlap. In the table, *Cost per overlay link* represents the total monitoring cost divided by the number of overlay links. The rows are sorted by increasing overlap coefficient. We observe that in general, the monitoring cost per overlay link decreases as overlap increases. However, the cost per link value for AS 1221 is slightly higher than that of AS 3257 although the former has a higher overlap coefficient. This may suggest that increasing overlap can only decrease the cost per link by a limited amount.

Table 3. Effect of link-level overlap on the lowest total monitoring cost

AS	Overlap coefficient	Lowest total cost	# of overlay links	Cost per link
3967	8.59	78	210	0.37
1755	9.21	92	272	0.34
6461	12.80	210	756	0.28
3257	17.08	222	992	0.22
1221	18.33	98	420	0.23

Percentage of overlay nodes. In this experiment, we vary the fraction of overlay nodes among all nodes in the network. We call this fraction *overlay node density*. We examine two Rocketfuel topologies using five different density values from 0.1 to 0.5, and random overlay node placement. Our ILP gives the results in Table 4 when $unitNativeCost = unitOverlayCost$. This result is consistent with the effect of link-level overlap. As the overlay node density increases, link-level overlap also increases, and the cost per overlay link decreases.

Table 4. Effect of overlay node density on the optimal monitoring solution

Overlay node density	Cost per link for AS 1755	Cost per link for AS 3967
0.1	0.75	0.71
0.2	0.34	0.37
0.3	0.25	0.28
0.4	0.15	0.17
0.5	0.12	0.13

5   Experimental Evaluation of Inference Errors

Composing an end-to-end measurement from other measurements can introduce an error in the result. We refer to this as *inference error*. One source of error may be packets traversing different sequences of router functions. For example, an end-to-end latency measurement probe may be forwarded along the fast path of a router, while probes that measure the latency of native links may be forwarded along the slow path. This makes the latter probe packets susceptible to processor delays, thereby introducing additional latency. Furthermore, some native link measurements may be inferred from overlay link measurements using arithmetic operations. This too introduces estimation error.

We represent the inference error for overlay links by computing the *absolute relative estimation error*. We compute this error value as a percentage:

$$\text{Abs. Rel. Est. Error Percentage}(e') = \frac{|\widehat{\rho}(e') - \rho(e')|}{\rho(e')} \times 100 \qquad (9)$$

where $\rho(e')$ is the actual measurement result for e' and $\widehat{\rho}(e')$ is the inferred result obtained through combining a different set of measurements.

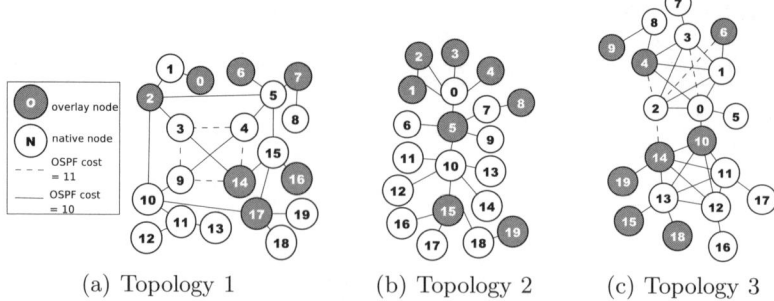

(a) Topology 1	(b) Topology 2	(c) Topology 3

Fig. 1. Three PlanetLab topologies we use. (a) represents a general AS topology. (b) has a tree-like structure which can be found on some campus-wide networks such as [2]. (b) can be interpreted as a graph of two interconnected ASes. Native links are assigned with different OSPF costs to avoid multiple shortest paths.

To assess the extent of inference errors, we conducted experiments on Planet-Lab [5] using three different overlay topologies shown in Fig. 1. We implemented these topologies as virtual networks on PlanetLab using PL-VINI, the VINI [1] prototype running on PlanetLab. In each experiment, we picked 20 PlanetLab nodes from different ASes as our native network and ran OSPF on this network with PL-VINI. Note that we cannot control the inter AS routing of these PlanetLab nodes. We treated the edges between these nodes on the PL-VINI network as native links. We picked 8 nodes out of the 20 as our overlay nodes, and assumed that these 8 nodes are fully connected to form an overlay network.

For each topology, we ran 4 rounds of measurements at different times. In each round, we measured the delay on all native and all overlay links by simultaneously running 100 pings on every link at a frequency of 1 per second. We calculated the delay from node a to node b as the average round-trip time over all ping results for native or overlay link $a - b$.

In order to find the optimal combination of links to monitor for these topologies, we ran our ILP on each of them with the objective of minimizing the total number of measurements. The output of the ILP gave us a set of overlay and native links to monitor. Using this output and the measurement results for the corresponding topology, we first inferred the measurements of the links that are not monitored, and then calculated the errors in these inferences using Eq. 9. The errors for all-native and basis set solutions are calculated in a similar manner.

Table 5 summarizes the results for all three topologies. The *Cost* column represents the lowest possible monitoring cost that can be achieved by each strategy. *Max* is the largest inference error observed in a certain strategy. Mn_i is the inference error averaged over all *inferred* overlay links, while Mn_a is the error averaged over *all* the overlay links in the network, with the difference being that direct overlay link measurements have no errors. Averaging over all overlay links does not reduce the error in the case of all-native monitoring because in this case all overlay links are inferred and none are measured directly. However,

Table 5. Costs and inference errors for different monitoring strategies

	Topology 1				Topology 2				Topology 3			
	$Cost$	Mn_i	Mn_a	Max	$Cost$	Mn_i	Mn_a	Max	$Cost$	Mn_i	Mn_a	Max
All-overlay	56	0	0	0	56	0	0	0	56	0	0	0
All-native	34	5.01	5.01	21.18	24	1.43	1.43	4.30	30	3.54	3.54	10.75
Basis set	38	2.68	0.86	20.29	26	0.96	0.51	2.79	26	1.13	0.61	4.95
Combination	26	3.43	2.70	20.12	18	1.58	1.35	3.17	24	2.35	1.68	10.75

$Mn_a < Mn_i$ in the basis set and lowest-cost combination strategies because some overlay links are directly measured and these zero errors bring down Mn_a.

Among the last three strategies, monitoring a combination of native and overlay links achieved the lowest cost, and monitoring a basis set of overlay links resulted in the smallest error. However, we should note that if we use a different cost definition, such as the total number of native links carrying probe traffic, these results may change significantly. For instance in topology 3, the last strategy uses a combination of 8 native and 16 overlay links, spanning a total of 42 native links , while the all-native solution spans 30 links and the basis set solution spans 52 native links. Our insight from these experiments suggests that in general, all-native solutions minimize bandwidth consumption, basis overlay set solutions minimize error, and using a combination of native and overlay links allows reducing the total number of measurements with comparable errors.

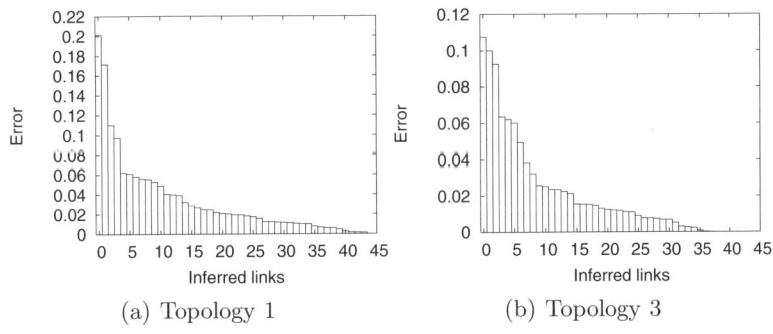

(a) Topology 1   (b) Topology 3

Fig. 2. Error rates of inferred overlay links

For the two topologies whose maximum errors are above 10%, we examine the error distribution among the inferred overlay links as shown in Fig. 2. We sort the inference errors from high to low and place them on the graphs from left to right. It can be seen that in both cases a few inferred links produce high errors that dominate the rest, increasing the mean error. If the ILP is aware of the overlay links that incur a high error when they are inferred, it can choose to monitor them directly and avoid these errors. Thus, adding certain error constraints to the ILP is a plausible step to improve its performance.

6 Conclusions

In this work we have proposed *multi-layer monitoring* as a flexible approach for overlay network measurement. We focused on the specific issue of determining the optimal mix of native and overlay link monitoring. We show that the overall cost of monitoring the network is the least when we allow native link measurements, as well as end-to-end measurements. We present a novel ILP formulation that when solved minimizes the cost of network monitoring with the appropriate combination of end-to-end and native network measurements. Through simulation studies, we observe that the optimal monitoring solution, i.e. the set of native and overlay links that minimizes the total monitoring cost while supplying sufficient information, depends on unit monitoring costs as well as the selection and placement of overlay nodes. We also find that the average monitoring cost per overlay link is lower for topologies where there is a high overlap between overlay links. Furthermore, we evaluate our approach through PlanetLab experiments with a focus on the question of *inference errors*.

Future work in this area should include: 1) applying our approach to multi-domain scenarios, 2) consideration of monitoring for metrics other than latency, 3) including error minimization as an objective in the optimization problem, 4) extending multi-layer monitoring to include Layer 2, 5) considering problems of dynamic monitoring which would allow changes in the monitoring mix over time in response to changing network conditions or changes in overlay topology.

References

1. Bavier, A., et al.: In VINI veritas: realistic and controlled network experimentation. In: Proceedings of ACM SIGCOMM, pp. 3–14 (2006)
2. CPR: Campus Wide Network Performance Monitoring and Recovery, http://www.rnoc.gatech.edu/cpr
3. GNU Linear Programming Kit (GLPK), http://www.gnu.org/software/glpk
4. Madhyastha, H.V., et al.: iPlane: An Information Plane for Distributed Services. In: OSDI, pp. 367–380 (2006)
5. Planetlab, http://www.planet-lab.org
6. Rocketfuel: An ISP Topology Mapping Engine, http://www.cs.washington.edu/research/networking/rocketfuel/
7. Seetharaman, S., Ammar, M.: Overlay-friendly Native Network: A Contradiction in Terms?. In: Proceedings of ACM HotNets-IV (November 2005)
8. Chen, Y., et al.: Algebra-based scalable overlay network monitoring: algorithms, evaluation, and applications. IEEE/ACM Trans. Netw. 15(5), 1084–1097 (2007)

Wireless

Understanding Channel and Interface Heterogeneity in Multi-channel Multi-radio Wireless Mesh Networks

Anand Prabhu Subramanian[1], Jing Cao[2], Chul Sung[1], and Samir R. Das[1]

[1] Stony Brook University, NY, USA
[2] Beihang University (BUAA), Beijing, China

Abstract. Multi-channel multi-radio architectures have been widely studied for 802.11-based wireless mesh networks to address the capacity problem due to wireless interference. They all utilize channel assignment algorithms that assume all channels and radio interfaces to be homogeneous. However, in practice, different channels exhibit different link qualities depending on the propagation environment for the same link. Different interfaces on the same node also exhibit link quality variations due to hardware differences and required antenna separations. We present a detailed measurement study of these variations using two mesh network testbeds in two different frequency bands – 802.11g in 2.4GHz band and 802.11a in 5GHz band. We show that the variations are significant and 'non-trivial' in the sense that the same channel does not perform well for all links in a network, or the same interface does not perform well for all interfaces it is paired up with for each link. We also show that using the channel-specific link quality information in a candidate channel assignment algorithm improves its performance more than 3 times on average.

1 Introduction

Wireless mesh networks based on commodity 802.11 radios are good vehicles to provide broadband network coverage at a low cost. Mesh networks, however, suffer from serious interference problems limiting their capacity due to broadcast nature of the wireless medium. A common method to improve capacity is to use multiple orthogonal channels that are already available in the 802.11 standard. The core idea is to limit the interference by using different channels for neighboring links. A network node can use multiple channels in two ways – either it dynamically switches channel on the radio interface for different transmissions, or it adopts a multi-radio solution, where each node has multiple radio interfaces tuned to different channels statically (or even dynamically, but at a longer time scale). Different links use different interfaces and thus different channels. The first method – dynamic channel switching on a single radio interface [2] – has proved practically hard as switching latency could be high in commodity

S.B. Moon et al. (Eds.): PAM 2009, LNCS 5448, pp. 89–98, 2009.
© Springer-Verlag Berlin Heidelberg 2009

802.11 radios [3]. Thus, the research community has pre-dominantly focused on the multi-radio solution.

The challenge in this case is to develop techniques for channel assignment, i.e., assigning channels to interfaces, subject to an appropriate optimization criterion, for example, reducing network interference or improving capacity. Since the number of interfaces in a network node is limited, this offers a constraint to the optimization problem. Many papers [7,9] (and references therein) have been published on this channel assignment problem, offering centralized or distributed solutions, investigating optimality questions, comparing performances, etc. *One singular limitation of all these works is that they all assume that the channels and radio interfaces are all homogeneous.* However in practice, the 802.11 channels vary significantly in Signal-to-Noise Ratio (SNR). Also, different radio interfaces on the same mesh nodes often provide different SNR measures even for the same channel. The goal of this work is to understand and demonstrate the heterogeneity in channels and interfaces via a set of careful measurements on two different wireless mesh network testbeds (802.11g and 802.11a) covering a wide-spectrum of possibilities. We show experimentally that the homogeneity assumptions often lead to very poor channel assignment. We followup the measurements with techniques to incorporate channel-specific link quality information in channel assignment algorithms to improve their performance.

The rest of the paper is organized as follows. In Section 2, we describe the details of our mesh testbeds. We present measurement results to understand channel heterogeneity in Section 3. Section 4 presents measurement results to understand interface heterogeneity in multi-radio mesh networks. We demonstrate how to improve the performance of channel assignment algorithms with channel heterogeneity information in Section 5. Related work is presented in Section 6 and we conclude the paper describing future directions in Section 7.

2   Testbeds

The measurements reported in this paper are from two different wireless mesh network testbeds (802.11g and 802.11a) set up in our departmental building as described below. The 802.11g testbed uses 10 Dell latitude D510 laptops each with one Atheros chipset based D-link DWL AG660 PCMCIA 802.11a/b/g card with an internal antenna. The transmit powers are fixed to 15 dBm and data rate to 11 Mbps. Measurements from this testbed were collected on 40 different links on three orthogonal channels 1, 6, 11 (2412, 2437 and 2462 MHz respectively) in the 802.11g band. The 802.11a testbed consists of 13 nodes each of which is a Soekris net4801 [1] single board computer (SBC). The PCI-slot in the SBC is expanded into 4 miniPCI slots using a PCI-to-miniPCI adapter. Four 802.11a/b/g miniPCI wireless cards based on Atheros chipset with external antennas are used in each mesh node. In order to overcome radio leakage problems, we physically separated the external antennas at a distance of about 0.5 meters based on measurements similar to [8]. Otherwise, there was a perceptible interference even among orthogonal channels across interfaces on the same node.[1] The transmit powers are fixed to 15 dBm and data rate to 6 Mbps. Measurements from this testbed were collected on 78 different links in 13 orthogonal channels (between 5180-5825 Mhz) in the 802.11a band. Note that the 802.11a testbed is relatively free from external interference as there are no other networks operating in this band in the building. However, there are indeed several 802.11g networks in our building. Their influence is impossible to eliminate. We, however, did our experiments in this network during late night and early morning when other active 802.11g clients are unlikely.

All nodes in both the testbeds run Linux (kernel 2.6.22 in laptops and kernel 2.4.29 in the Soekris boxes) and the widely used `madwifi` device driver (version v0.9.4) for the 802.11 interfaces. We used standard linux tools such as `iperf` to send UDP packets on the sender node for each link measured and `tcpdump` on the receiver node running on a raw monitoring interface to capture the packets. This gives us the additional prism monitoring header information such as the received signal strength (RSS), noise, channel and data rate for every received packet.

3   Channel Diversity

This section shows the results of our measurement study to understand the heterogeneity in channels due to varying path loss of different frequency bands. In the following, we first show that Received Signal Strength (RSS) of packets in each link is relatively stable in each channel and is a 'good' metric to compare the performance of any given link when using different channels.

3.1   Long Term Variation of RSS

We study a single link in the 802.11a testbed for a *24 hour* period by sending 1000-byte UDP packets at a rate of 100 packets per second. We repeat this experiment on 7 different 802.11a channels for the same link. Figure 1(a) shows the Allan deviation in the RSS values in each of the 7 channels at different time intervals ranging from 100 ms to 10 hours. Allan deviation is used as a metric to quantify the burstiness of variation in any quantity. The median variation is about 1.5 dBm and the 90% variation is about 2.5 dBm in a single channel. The variations are similar across all 7 channels. We see that the variation at different intervals are small considering the minimum granularity of RSS measurements is 1 dBm. This figure shows that in *any given channel*, the variation in RSS value is minimal and sampling RSS values at smaller intervals (in the order of tens of seconds) can be representative of longer measurements. We also see similar results in the 802.11g testbed which are not reported here due to space constraints.

---

[1] Even with this setup, we could use only a subset of orthogonal channels without interference. These are 7 channels (channels 36, 44, 52, 60, 149, 157, 165) out of possible 13 orthogonal channels. Thus, we used these 7 channels for channel assignment in Section 5. However, we used all 13 channels to study the channel characteristics in Sections 3 and 4.

3.2   Relation between RSS and Delivery Ratio

Now that we have seen that RSS is relatively stable over long periods of time, next our goal is to show that RSS is a good predictor of link performance in each channel. For this, we studied 78 different links in the 802.11a testbed by sending back-to-back 1000-byte packets in each link using the 13 orthogonal channels for a period of 60 seconds one after another and measured the average RSS value and delivery ratio for each link in different channels. Figure 1(b) shows the relationship between average RSS and the delivery ratio of the links in our 802.11a testbeds. It shows a scatter plot of average RSS vs. delivery ratio of each link for all channels. The interpolations (the dark lines) of the aggregated data are also shown. Visually it appears that the RSS vs. delivery ratio statistics is independent of channels – no definite channel specific pattern emerges. We have also computed the R^2 value for each individual channel data with respect to the interpolation (noted in the plots). The R^2 values are similar across channels - varying between 0.82–0.94. This shows that RSS is a good predictor of delivery ratio and this relationship is relatively independent of the channel used. Note that delivery ratio (or, throughput) is a commonly accepted performance metric for the upper layer protocols. We observed similar characteristics from measurements in the 802.11g testbed. Thus, we can focus on RSS alone to understand channel and interface specific behavior as this fundamental metric is influenced by the *propagation environment*.

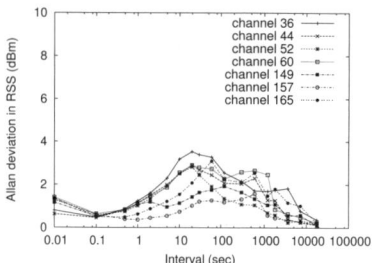
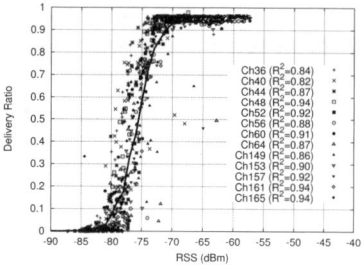

(a) Long term variation of RSS values for a single link in 7 different 802.11a channels.

(b) Relationship between average RSS value and delivery ratio in different channels in our 802.11a testbed.

Fig. 1. Characteristics of RSS metric

3.3   Link Behavior in Different Channels

Now we look at the average RSS value (with 95% confidence interval) on each channel for two sample links in each testbed. See Figure 2. Figures 2(a) and 2(b) show the performance of two 802.11g links. In both cases, we see considerable variation in RSS in different channels. In the first case, even though there is variation in RSS, the delivery ratios do not vary much. This is because the RSS values are already quite high. In the second case, we see that the delivery ratio of the link is good in channel 1 and 6 but is quite poor in channel 11. A

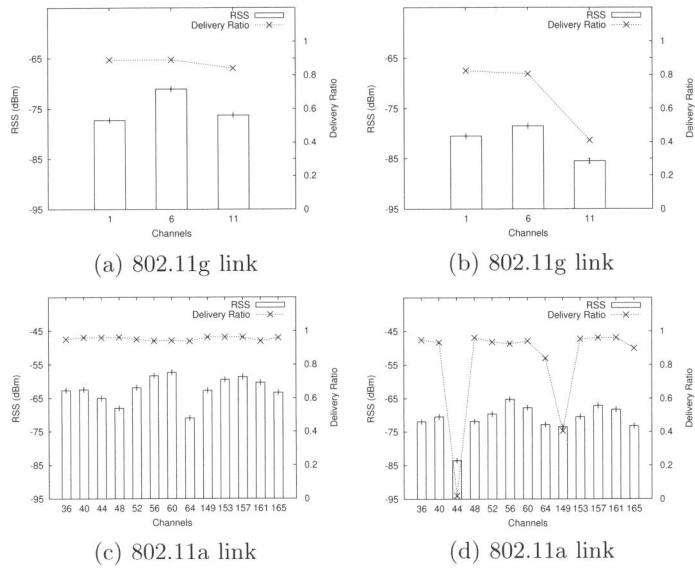

(a) 802.11g link

(b) 802.11g link

(c) 802.11a link

(d) 802.11a link

Fig. 2. Variation of RSS and delivery ratio using different channels on sample links in our two testbeds

similar behavior is observed in the 802.11a testbed. See Figures 2(c) and 2(d) for two sample links. These results demonstrate that RSS on a link could be channel-specific and this can impact the delivery ratio significantly.

It is now interesting to study how much variation is there in RSS values for each of the 40 links in the 802.11g testbed and 78 links in the 802.11a testbed. In Figure 3(a) we show the range of variation in RSS value for each link in the 802.11g testbed. The bars show the maximum and minimum RSS value for each link considering all channels. The median RSS range (i.e., the median of the differences between the maximum and minimum over all links) is about 6 dBm and the 90-percentile RSS range is about 12 dBm. Figure 3(b) shows the RSS variation in the 802.11a testbed. In this case, the median RSS range is about 11 dBm and the 90-percentile RSS range is about 18 dBm. This is significantly higher than the variation of RSS in a single channel as noted previously. *Evidently, there are considerable variations in RSS values across channels.* The variation in the 802.11a testbed is higher. This is because the path loss characteristics are frequency specific and the 802.11a band (5180-5825MHz) is much wider compared to the 802.11g band (2412-2462MHz).

In both the plots, the horizontal arrow shows the RSS threshold values. Note that many links the RSS range crosses the threshold indicating *such links perform poorly in some channels, while performing quite well in some others.*

Now, it will be interesting to find out whether there is any one channel that is good for all links. In Figure 3(c) and 3(d), we show how many times each channel is the best based on the RSS values considering all links studied. We see that in both testbeds, there is no clear winner among channels. Each link performs differently in different channels. The RSS values are not correlated with

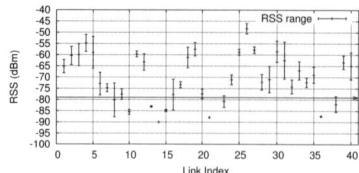

(a) Range of RSS variation in each link in 802.11g testbed across all 3 orthogonal channels.

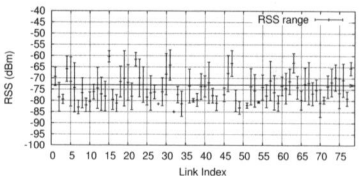

(b) Range of RSS variation in each in the 802.11a testbed across all 13 orthogonal channels.

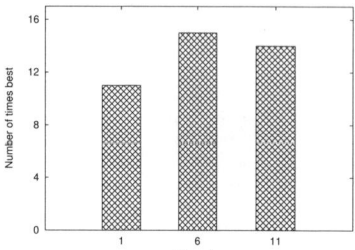

(c) Number of times each channel is best based on the RSS values on each link in the 802.11g testbed.

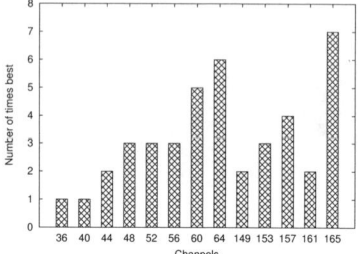

(d) Number of times each channel is best based on the RSS values on each link in the 802.11a testbed.

Fig. 3. Link behavior across different channels in the two testbeds

the channel frequency. If this was the case, the channel 36 in the 802.11a band and channel 1 in the 802.11g band should have the best RSS values in all links. Some channels do exhibit better overall performance relative to their peers (e.g., channels 165 and 64 for 802.11a testbed). But generally speaking, any channel could be the best for some link. *This makes it impossible to judge which channels to use for a given link without doing actual measurements on the links.*

4   Interface Diversity

For a given link between two multi-radio nodes, the choice of actual radio interfaces to use for this link could impact the link performance. The reason for this is two fold. First, there could be inherent manufacturing variations between the interfaces even though they use the same card model. Second, the antennas for the interfaces need to be situated at a distance to prevent radio leakage issues so that the orthogonal channels do remain orthogonal in practice [8]. This makes the actual distance between different antenna pairs for the same node pair slightly different (noted in Section 2). This issue is more significant in 802.11a as it provides shorter ranges relative to 802.11g. On the other hand, 802.11a is indeed attractive for multichannel work, as it provides many more orthogonal channels.

To understand the variations caused by interface selection, we study 20 links (a subset of the 78 links studied before) in our 802.11a testbed using 16 possible interface pairs for each link. We select the same channel (channel 64, one of the

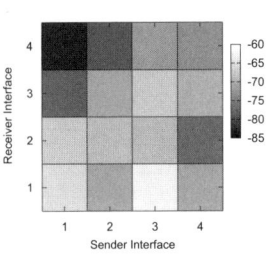

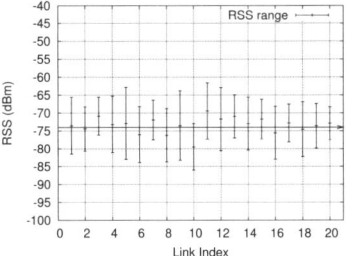

(a) RSS values (in dBm) for 16 possible interface pair combinations on a sample link.

(b) Range of RSS value between different interface pair combinations for each link.

Fig. 4. Interface heterogeneity in multi-radio nodes in 802.11a testbed

good performing channels) for this measurement on all links in order to isolate the effect of interface selection.

Figure 4(a) shows the RSS values on all 16 possible interface pair combinations for a sample link. Here we see that the RSS value varies between -60 dBm to -85 dBm. Considering the RSS threshold (about -74 dBm), the link shown here has a very poor delivery ratio when certain interfaces are used (e.g., 1 to 4). However, some other interfaces would have a good delivery ratio (e.g., 3 to 1). *It is also interesting to note that we cannot say that a specific interface has poor performance.* For example, if we consider the interface 1 on the sender node, it has varying performance based on the receiver interface.

In Figure 4(b), we show the range of variation in RSS values between the 16 possible interface combinations for each of the 20 links studied. Each bar shows the maximum and minimum RSS value for each link considering all 16 combinations. Note the significant variation in RSS values among different interface pairs. The median and 90-percentile RSS variation is about 12 dBm and 16 dBm respectively. Also note that most of these ranges straddle the RSS threshold (-74 dBm). This means the delivery performance can indeed significantly vary depending on the interface choices. *A channel assignment algorithm unaware of such variations can easily choose a bad interface pair for a link even though there are better interface pairs that could be potentially used.*

5   Channel Assignment Algorithm

In this section, we demonstrate the potential of using channel-specific link quality information in existing channel assignment algorithms to get better performance. For this purpose, we modify the greedy channel assignment algorithm proposed in [9] to use the channel-specific link quality information when assigning

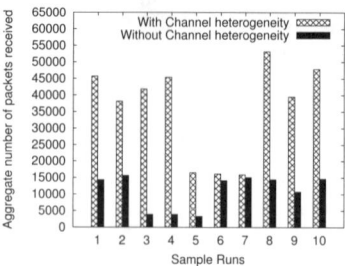

Fig. 5. Aggregate number of packets received when a set of 10 links transmit packets simultaneously. Each sample run consists of a different set of 10 links in the network.

channels for links. The greedy channel assignment algorithm assigns channels to *links*[2] in a greedy fashion trying to minimize the overall interference in the network. At the same time it satisfies the interface constraint, i.e., ensures that the number of channels assigned to links incident on a node does not exceed the number of interfaces on the node.

The channel assignment algorithm works as follows: Initially, none of the links are assigned channels. The algorithm iterates through each link that is not assigned a channel yet and chooses a feasible set of channels that obey the interface constraint. From this feasible set of channels, it selects a channel that minimizes the overall *network interference* which is modeled using a conflict graph. The algorithm terminates when no further assignment of channels to links can reduce the network interference. Note that among the channels in the feasible set, it is often the case that more than one channel can lead to the minimum interference. Since the algorithm is unaware of possible difference in link quality in different in channels, it chooses one channel arbitrarily. *Note that this is a singular limitation in all channel assignment algorithms in current literature as they do not use channel specific link quality information to make a choice.* In the new version of the greedy channel assignment algorithm, we use the channel-specific link quality information (e.g. RSS on different channels) to make this choice. Given RSS values are relatively stable, short term measurements (one time or periodic) are good enough to estimate the link quality in different channels. These measurements can be done whenever the channel assignments are recomputed. Estimating the periodicity of channel assignment depending on the environment and channel conditions is one of our future work.

In our 802.11a multi-radio testbed, we use 7 orthogonal channels (channels 36, 44, 52, 60, 149, 157, 165) and 4 interfaces in each node to study the performance of the channel assignment algorithm. In Figure 5, we show the performance of the greedy channel assignment algorithm with and without the channel-specific link quality information. We used periodic probes sent at 100 packets per second in each channel for 1 second to measure the link quality in different channels

---

[2] Since it assigns channels to links directly, it is difficult (but not impossible) to incorporate the interface-specific information in this algorithm. We consider exploring the use of interface-specific information as a part of our future work.

on each link before running the greedy algorithm that uses channel-specific link quality information. The horizontal axis shows 10 different experimental runs. In each run, we send back-to-back UDP packets on 10 randomly chosen links simultaneously. The two versions of the channel assignment are used to assign channels for these 10 links. For each channel assignment, the experiment is run for 60 seconds and the aggregate number of packets received is measured. Note that the channel assignment algorithm using the channel-specific link quality information performs very well in all experimental runs compared to the case when all channels are considered homogeneous. Except in two cases (runs 6 and 7), the improvements are quite substantial - varying between 2-8 times. We noted that in the two cases where performance improvements are marginal, use of channel-specific information did not result in a very different channel assignment. Overall, the average improvement was by a factor of about 3.

6   Related Work

There is a growing body of literature that use multiple channels to reduce interference in wireless mesh networks [2,9,7]. Many of them use multi-radio solutions [6,9,7] (and references therein) to eliminate the need for dynamic channel switching. However, none of these works consider the variations in link quality depending on the channel or interface chosen for communication. Channels are always assumed to be homogeneous and link quality to be independent of interface selection or choice of channel.

Recently, Das et al [4] have observed variation in routing metrics in different channels in wireless mesh networks. However, their work primarily focuses on comparing different routing metrics and understanding their dynamics. In [5], the author has observed variation in link quality in multiple channels when studying interference maps in 802.11 networks. The paper studied one 802.11a link and showed variation in delivery ratio in different channels. Our work quantifies the variation in using different channels and interface pairs using extensive measurements in two different mesh testbeds operating 802.11g and 802.11a bands and using different hardware platforms. We also show that the variations in link quality are not correlated to frequency of the channels. We also experimentally demonstrate that utilizing channel and interface-specific information in channel assignment algorithms improves performance significantly.

7   Conclusions

This paper presents a detailed measurement study of channel and interface heterogeneity in multi-radio wireless mesh networks using measurements from two mesh testbeds using different hardware platforms and frequency bands (2.4GHz for 802.11g and 5GHz for 802.11a). We quantify the variation in link quality when using different channels and interface pairs and show that choosing the right channel and interfaces for a link can improve its performances significantly. We also demonstrate that this variation is 'non-trivial' in the sense that same channel does not perform uniformly well for all links, or the same interface does not perform uniformly well for all other interfaces it is paired up with.

All prior channel assignment works in literature ignore this important assumption. We demonstrate how the channel heterogeneity information can be incorporated in an existing channel assignment algorithm to improve its performance. An important future direction of our work is to develop methods to measure these variations efficiently, understand how often they need to be repeated and design channel assignment schemes that take both channel and interface variations into account and come up with efficient solutions.

References

1. Soekris Engineering, http://www.soekris.com/
2. Bahl, P., Chandra, R., Dunagan, J.: SSCH: Slotted seeded channel hopping for capacity improvement in IEEE 802.11 ad-hoc wireless networks. In: MOBICOM (2004)
3. Chandra, R., Bahl, P., Bahl, P.: MultiNet: Connecting to multiple IEEE 802.11 networks using a single wireless card. In: INFOCOM (2004)
4. Das, S.M., Pucha, H., Papagiannaki, K., Hu, Y.C.: Studying Wireless Routing Link Metric Dynamics. In: IMC (2007)
5. Niculescu, D.: Interference Map for 802.11 Networks. In: IMC (2007)
6. Ramachandran, K., Belding, E., Almeroth, K., Buddhikot, M.: Interference-aware channel assignment in multi-radio wireless mesh networks. In: INFOCOM (2006)
7. Raniwala, R., Chiueh, T.: Architechture and algorithms for an IEEE 802.11-based multi-channel wireless mesh network. In: INFOCOM (2005)
8. Robinson, J., Papagiannaki, K., Diot, C., Guo, X., Krishnamurthy, L.: Experimenting with a Multi-Radio Mesh Networking Testbed. In: WiNMee Workshop (2005)
9. Subramanian, A.P., Gupta, H., Das, S.R., Cao, J.: Minimum Interference Channel Assignment in Multi-Radio Wireless Mesh Networks. IEEE Transactions on Mobile Computing 7(11) (2008)

Access Point Localization Using Local Signal Strength Gradient

Dongsu Han[1], David G. Andersen[1], Michael Kaminsky[2],
Konstantina Papagiannaki[2], and Srinivasan Seshan[1]

[1] Carnegie Mellon University
[2] Intel Research Pittsburgh

Abstract. Many previous studies have examined the placement of access points (APs) to improve the community's understanding of the deployment and behavioral characteristics of wireless networks. A key implicit assumption in these studies is that one can estimate the AP location accurately from wardriving-like measurements. However, existing localization algorithms exhibit high error because they over-simplify the complex nature of signal propagation. In this work, we propose a novel approach that localizes APs using directional information derived from local signal strength variations. Our algorithm only uses signal strength information, and improves localization accuracy over existing techniques. Furthermore, the algorithm is robust to the sampling biases and non-uniform shadowing, which are common in wardriving measurements.

1   Introduction

Locating the source of a radio frequency (RF) transmission is important for a wide range of purposes. These include finding rogue access points (APs), creating wardriving maps, and estimating the RF propagation properties of an area. The traditional approach to localizing an AP is to perform wardriving like measurements (i.e., measure received signal strength and location information) and to apply a number of common techniques to process the data. Unfortunately, the types of algorithms that can be applied to such data are limited, and state-of-the-art algorithms exhibit high error and high variation. Recent studies [1] have shown that using erroneous results produced by state-of-art localization algorithms impair the performance of mobile user positioning systems like Place Lab [2] and cause inaccurate estimates of coverage and interference.

One approach to improving localization accuracy is to use more sophisticated data collection techniques. For example, Subramanian et al. [3] recently improved accuracy by using angle of arrival (AoA) information collected with a steerable beam directional antenna. While such techniques improve accuracy, the cost of hardware and human time is high. In this paper, we present a novel AP localization algorithm called *gradient* that uses only information collected from conventional wardriving, and that does not require extra hardware. Our approach uses the local signal strength distribution to estimate the direction of the AP. The gradient algorithm localizes the AP by combining directional estimates from

S.B. Moon et al. (Eds.): PAM 2009, LNCS 5448, pp. 99–108, 2009.

multiple vantage points. We show that *gradient* improves the mean accuracy by 12% over the state-of-the-art algorithm, and reduces the maximum error and standard deviation of errors by more than 33%. This paper describes the key insights and the design of the *gradient algorithm*, verifies the idea through simulation (Section 2), and compares its real-world performance against existing methods (Section 3).

2   Localization Algorithm

The goal of our work is to estimate the location of an AP given a set of received signal strength (RSS) measurements from different locations. These RSS measurements are typically obtained by passively monitoring 802.11 frames from a moving vehicle—a practice known as *wardriving*. The data contains measurement locations (x, y) and a received signal strength RSS at each location.

We first examine the the most commonly used localization algorithms, centroid, weighted centroid [2] and trilateration [4], to motivate the need for a better algorithm. Given the set of measurement points $<x_i, y_i, RSS_i>$ of an AP, centroid algorithms locate the AP at the averaged location: $\left(\sum_{i=1}^{N} w_i x_i, \sum_{i=1}^{N} w_i y_i \right)$.

For centroid, $w_i = 1/N$, and for weighted centroid, $w_i = \dfrac{SNR_i}{\sum_{j=1}^{N} SNR_j}$.

Trilateration [4] estimates the distance from the signal source at a measurement point using the RSS and combines these distance estimates to infer the location of the AP. The RSS is converted to distance using the log-distance path loss model [5]. The model defines the path loss $(\overline{PL})$ from transmitter to receiver as a function of distance (d) as $\overline{PL}(\text{dB}) = \overline{PL}(d_0) + 10n \log \left(\frac{d}{d_0} \right)$, where d_0 is a reference distance and n is the path loss exponent.

Despite their simplicity, centroid-based algorithms perform as well as other existing algorithms that use signal strength [1, 3]. To understand why our gradient algorithm can outperform all of these algorithms, we first examine one representative case where centroid algorithms perform poorly.

2.1   Motivating Example

Figure 1 shows wardriving measurements collected for one AP inside an apartment. The building is near a three-way junction. The AP is located towards the front of the building. Each measurement point is shaded to indicate the received signal strength according to the gray-scale legend on the right side of the figure.

Because the measurements are taken from a car, the path of the measurement points reflects the shape of the roads. *Wardriving measurements introduce strong sampling biases.* Figure 1 also shows non-uniform signal propagation. The area in front of the AP has denser and stronger (light-colored) measurements compared to the area behind the building. This is because the signal behind the building is shadowed by multiple walls, while the front of the building has fewer obstructions. Such non-uniform shadowing is also typical in wardriving measurements.

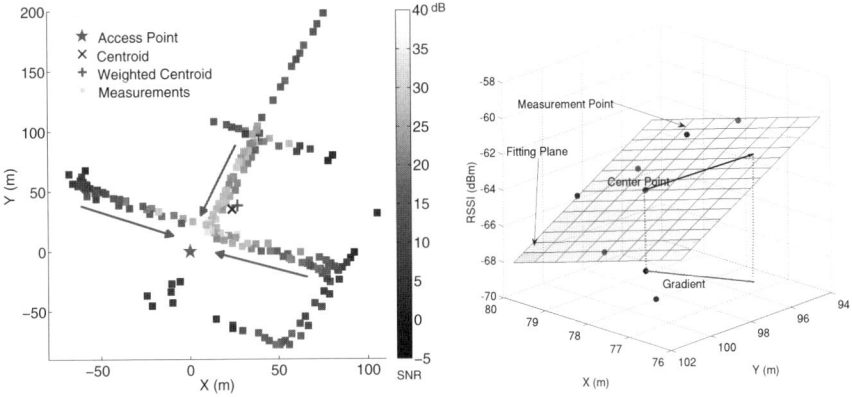

Fig. 1. Real world measurement of an AP located at the origin

Fig. 2. Drawing the arrow

As a result, both centroid and weighted centroid give high errors of 51.8m and 52.3m (Figure 1). Centroid algorithms suffer when subject to 1) biased sampling, which is exacerbated by the layout of the road, and 2) non-uniform signal propagation due to different levels of shadowing.

Despite the skewed distribution, the signal strength distribution remains a useful hint in localizing the AP. We see a clear trend of increasing signal strength towards the AP, as denoted by the arrows in Figure 1. These arrows provide useful information about the location of AP. Based on this observation, we develop a novel AP location algorithm which we call the *gradient algorithm*.

2.2   Gradient Algorithm

The main idea in the gradient algorithm is to systematically point arrows towards the signal source. We estimate the direction of an AP from every measurement point by calculating the direction of strongest signal in the neighboring area. After this process, each measurement point has an *arrow* that points to the direction of AP. By *combining* the resulting arrows, we estimate the location of the AP. While individual arrows have some error, overall these would cancel out during this stage.

The algorithm's effectiveness arises from two properties of the signal propagation: First, despite the inherent noise in the measurement, distance from the source AP remains one of the most important determinants of RSS. Second, obstructions cause sudden attenuation of signal that weakens the correlation between distance and RSS in the global space. However, measurements near-by in space tend to have traveled through the same obstructions. Thus, when considering only local signal strength, distance is the single most important factor. The algorithm leverages these observations and estimates the direction of the AP by finding the direction of strongest signal in near-by space.

Although the effect of large-scale non-uniform shadowing may be constrained in a local area, other small-scale factors — muti-path propagation, reflection, and errors in GPS positioning, among others— affect RSS. The gradient algorithm treats them as noise and relies on averaging to mitigate their effect.

These steps of the gradient algorithm operate as follows:

Preprocessing. This phase averages signal observations to mitigate the effect of various sources of error. First, we embed the recorded GPS coordinate into a two dimensional coordinate system. Then we place a unit meter grid on the map and average the RSS and coordinates that fall into the same grid square.

Drawing the arrows. We assume free-space propagation locally, and find the direction of the AP by calculating the direction in which the RSS increases the most. We do so by fitting a plane to "near-by" measurements in the x-y-RSS space, and taking the gradient of the plane. We use minimum mean-square error (MMSE) fit to overlay a plane, and take the gradient at the center point as the direction of the AP in the x-y plane (Figure 2). We call this gradient an *arrow*. We locate "near-by" space using a square centered at the center point, and define "window size" to be half the length of the side of the square.

The window size is critical because it determines the area in which the signal propagation is approximated as free-space. The optimal value depends on the density of measurement points and their spatial distribution. Therefore, there is no one-size-fits-all value across different APs. We revisit this problem at the end of the section.

Combining the arrows. Once we determine the direction of the AP from each measurement point by drawing arrows as described above, we locate the AP at the position to which all of the arrows point. We position the AP at the location that minimizes the sum-squared angular error from the arrows. In calculating the sum-squared error, we weigh each error by the SNR at the measurement point. This step is similar to the technique used by Subramanian et al [3], as well as other algorithms that use AoA.

Any standard optimization tool can be used for this minimization. However, most tools do not guarantee a global minimum and they require a good initial guess from which they start searching. Because the performance of any particular optimization algorithm is not of interest, we exhaustively search the space every meter, limiting it to a 100m x 100m square centered at the maximum RSS point. When using an optimization tool, we recommend using the maximum RSS point as the initial guess.

Window size. The window size is important in accurately predicting the direction of the AP from each measurement point. If the window is too small and contains few measurements, the resulting arrow is likely to have a high error. On the other hand, making the window too large is not desirable because it breaks our free-space assumption.

Because it is difficult to predict the best value for the window size at each measurement point, we reduce the problem to finding a single optimal value of the window size for each AP. We use the following heuristic to determine this value. First, we produce a series of estimated locations using window sizes from

1m to 50m. If the area covered by a window contains at least four measurements, we use that area to draw an arrow. At the end of the arrow drawing phase, if the arrows are drawn for fewer than 30% of all measurement points, the we do not produce an estimate for the window size.

The gradient algorithm starts using a window size of 1m and repeatedly increases it by one until five consecutive estimates converge within 5m from their average. At that point the algorithm terminates and reports the average estimate for the AP location. In Section 2.3, we verify the heuristic using simulation.

2.3   Simulation

We use simulation to verify the basic idea behind the gradient algorithm, and show that it is accurate and robust to sampling bias and non-uniform shadowing.

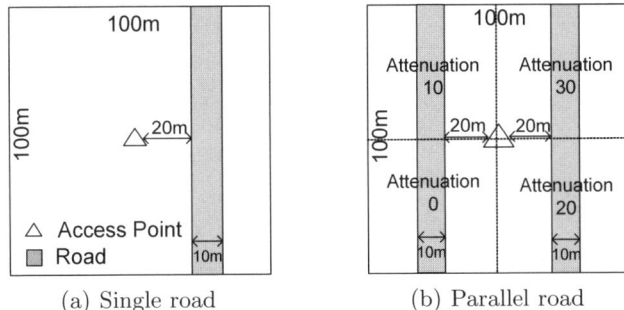

(a) Single road   (b) Parallel road

Fig. 3. Simulation topologies: measurements are generated only on the road

Accuracy. We create a single and a parallel road topology (Figure 3) to verify the performance of the algorithm. We place an AP at the center of the maps and generate measurement points randomly along the road. The RSS at each point is calculated using the log distance path loss model of Section 2. We set the reference distance d_0 to 1m, and subtract the path loss from signal strength at the reference point to get the RSS. For the Figure 3(a) topology, we vary the number of measurement points from 10 to 50. For (b), we generate 200 points and run another simulation that further attenuates the signal strength by the number on each quadrant to simulate non-uniform shading. Table 1 shows the average error of 20 runs using the gradient and centroid algorithms.

Gradient performs better as the density of measurement increases, and localizes the AP within 1 meter when the number of measurements exceeds 35. In contrast, the centroid algorithms suffer high error due to the biased sampling. When only non-uniform shading is introduced, gradient suffers much less than the weighted centroid, but in general needs denser measurements.

To see the effect of the path loss exponent, we vary the value of n from 2.0 to 5.0 which is the typical range in outdoor environments [5]. Figure 4 shows the average, minimum and maximum estimation error for the gradient algorithm using 10 different sampled topologies for Figure 3(a). The gradient algorithm appears unaffected by the value of the path loss exponent.

Table 1. Localization error for the topology in Figure 3(a) and (b). Units in meters.

Number of measurements					
10	15	20	25	30	35
Gradient 10.0	6.5	1.8	1.2	1.1	1.0
Centroid 26.4	26.4	26.3	25.8	25.5	25.8
W. Cen. 26.1	26.2	26.1	25.7	25.3	25.6

With non-uniform attenuation	
Gradient 6.7 W. Cen.	16.0
Without attenuation	
Gradient 0.3 W. Cen.	3.0

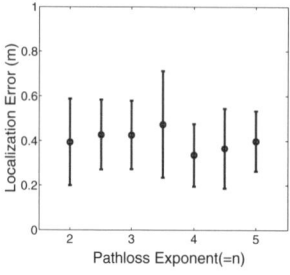

Fig. 4. Error versus path loss exponent

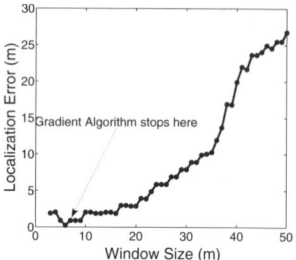

Fig. 5. Error versus window size

Effect of window size. As mentioned in Section 2.2, the window size is important for accuracy. To understand the effect, we generate 50 measurement points using Figure 3(a)'s topology and produce a series of estimates while varying the window size from 1 to 50m. The resulting curve (Figure 5) features a natural minimization point, where the gradient algorithm stops.

When the window is small, only a few measurement points are considered when drawing the arrow, thus leading to high error. As the window grows, our two assumptions break down. The free-space assumption stops holding because the large window now contains effects of non-uniform shadowing. (This effect is not present in the free-space simulation.) Second, because the global signal strength distribution is not linear, the distribution inside the large window also becomes non-linear. This results in errors due to underfitting since the gradient algorithm attempts to apply a linear fir to a non-linear distribution.

3 Evaluation

In this section, we evaluate the performance of the gradient algorithm in real environments. We first collect measurements and use them to compare the algorithm's accuracy with that of previously proposed approaches. We demonstrate that the algorithm is robust and clearly outperforms distance estimation based on signal strength.

3.1 Data Collection Framework

We collected wardriving measurements using the following setup:

Hardware and software. We used Intel-based laptops running Linux, with a Pharos GPS device, an Atheros wireless cardbus card, and an external omnidirectional antenna mounted on top of a car. In order to capture more packets, we used two laptops, with one scanning only the popular orthogonal channels 1, 6 and 11, and the other scanning channels 1 through 11. The madwifi driver and *kismet* were used to capture packets in monitor mode. The madwifi driver reports continuous received signal strength levels, which is important to our algorithm. *Kismet* records GPS coordinates, all received packets and their signal strength. From the *kismet* log, we extract an <AP MAC address, SSID, latitude, longitude, SNR, noise> tuple.

Wardriving. We took measurements in the residential Pittsburgh neighborhood of Squirrel Hill. The area consists mostly of detached homes, townhouses and small apartments. We drove at about 15mph, scanning each side of the road two to four times. Slow speed and multiple scans allowed us to get more measurements per AP. Scanning both sides of the road increases the number of measurements inside a small area, which improves the accuracy of our algorithm.

Ground truth. Determining the actual location of an AP is crucial for evaluation, but is difficult, especially in the unplanned deployment setting that we examine. We obtained ground truth locations in two ways. First, we knew some of the AP owners personally. Second, we observed that many SSIDs contained either street address or family names. Using an online phone book [6] plus the local zipcode, we mapped family names to addresses. We further verified that the resulting address was reasonably close to the observed location for the AP.

For each candidate ground truth AP, we measured the actual GPS coordinates on the street in front of the addressed building. This step was surprisingly important— address to GPS coordinate conversion services such as Google Maps often produce inaccurate mapping that were off by a few houses. We manually processed the measured coordinates by positioning the AP at the center of the house using Google Maps. This process located twenty-five AP's.

3.2   Real World Experiment

We compare the accuracy of the gradient algorithm with that of centroid and trilateration using the data we collected in Section 3.1. We also compare with an algorithm that we developed called cone-fit. Cone-fit uses the same path loss model of Section 2. Given the path loss exponent as a parameter, it estimates, for all measurement $<x_i, y_i, RSS_i>$, the RSS from the path loss model and produces $<x_i, y_i, EstimatedRSS_i>$ by placing the AP at a location. It locates the AP at the location where $\sum (EstimatedRSS_i - RSS_i)^2$ is minimized. Visually, it is fitting a 3 dimensional cone to $<x, y, RSS>$ space whose skirt is shaped by the path loss exponent.

Note that unlike gradient, trilateration and cone-fit are parameterized. Trilateration takes as inputs the path loss exponent and the signal strength at the reference point, and cone-fit takes the path loss exponent as input. We give them

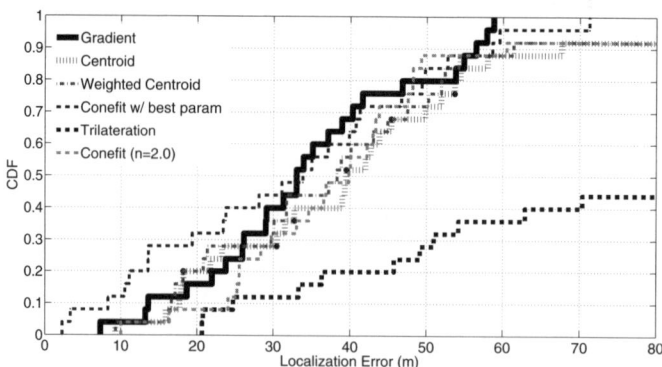

Fig. 6. Comparison of gradient, trilateration, centroid, and cone-fit

full advantage by calculating the best parameter given the actual known location of each AP. Figure 6 shows the localization error of these algorithms.

The mean error and standard deviation of gradient were 34m and 14m respectively. Gradient improves the mean error by 12% over weighted centroid, and performs better than centroids and trilateration. Moreover, it has the lowest worst case error, giving a factor of 1.4 improvement in standard deviation over weighted centroid. The maximum error of gradient was 59m versus weighted centroid's 88m. Cone-fit's average is 7% better than gradient only when the best path loss exponent was given from the actual AP's location, but when the parameter is fixed (e.g., n=2.0), it was worse than gradient. Its performance was highly variable depending on the exponent, and degrades under non-uniform shadowing and biased sampling.

Case study. To understand the characteristics of the gradient algorithm, we describe some example scenarios from the actual measurement (Figure 7). The figures show arrows at measurement points and the localization results of various algorithms[1]. Figures 7(a),7(b), and 7(c) show some skewed distributions that frequently occur in wardriving measurements. Note that a cut through the AP can be made in the x-y plane such that one side of the cut contains the vast majority of measurement points. Gradient performs well in these cases, unlike other algorithms. The more uniform distribution in Figure 7(d) provided the smallest error (of all 25 APs) for weighted centroid, because the AP was located at the corner of a junction. All algorithms perform relatively well in this scenario. While all algorithms perform well in under a uniform distribution of measurement points, only gradient performs well in all shown cases.

4   Related Work and Discussion

Localization systems fall into three rough categories based upon the information they use [4]: received signal strength, time-based, and angle-of-arrival (AoA).

---

[1] Trilateration is not shown because its errors were too large.

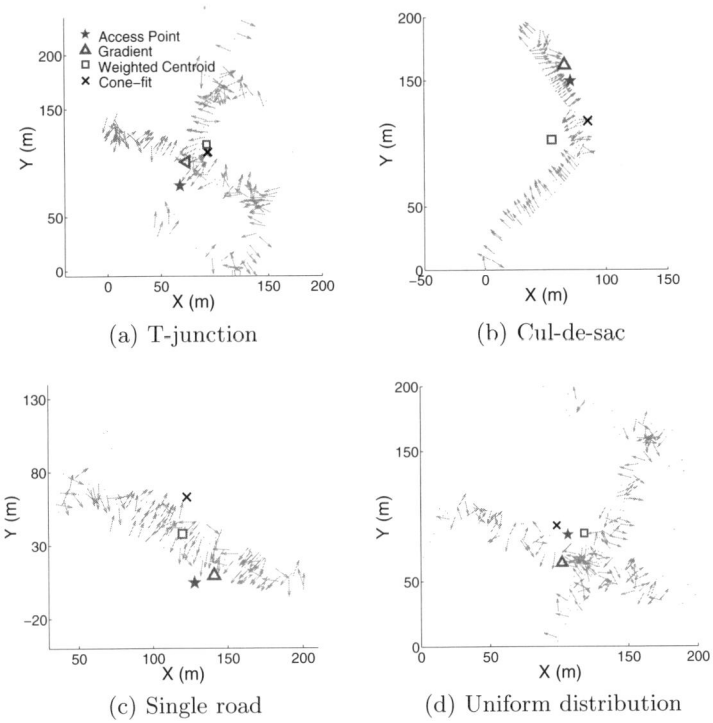

(a) T-junction   (b) Cul-de-sac

(c) Single road   (d) Uniform distribution

Fig. 7. Case study. Arrows are trimmed to improve readability.

RADAR [7] uses RSS to infer the distance between the signal source and the receiver. However, a site-survey is required to compensate for the non-uniform shading. Similarly, time-based approaches [8, 9] require accurate time synchronization or use an additional medium such as ultrasound to address timing issues.

AoA systems measure the incident angle of the signal at the receiver. Subramanian et al. [3] proposed an AP localization algorithm that uses AoA information measured with a steerable beam directional antenna. This state-of-the-art algorithm significantly reduces the localization error. However, the evaluation is based on APs that were modified to transmit a large number of packets and placed on the same channel to speed up the data collection. Simulations of normal-speed data collection, commensurate to a real-world situation, showed that the median localization error was comparable to that of gradient.

Of the three types of information, RSS is the only information collected in conventional wardriving measurements. Furthermore, few techniques used in user localization systems can be applied to AP localization in ad-hoc environments. Therefore, centroid and weighted centroid have been commonly used [2, 1]. Our work takes a novel approach in that we measure only RSS, but derive AoA equivalent information.

Finally, particle filters [10] provide a new approach to combine information obtained from each measurement regardless of its type. It provides a probabilistic mechanism to combine data instead of the deterministic ones used by most localization techniques. An interesting avenue of future work would be to combine our work with a particle filter, to estimate locations using both the RSS and the derived AoA information from our work.

5   Conclusion

This paper presents a novel access point localization algorithm called *gradient* and its performance evaluation. Gradient is based on the idea that the signal strength indirectly reflects the direction where the signal comes from. The algorithm does not require extra hardware, and data can be collected by normal wardriving.

Our evaluation shows that gradient overcomes the sampling bias that is inherent in wardriving measurements. Furthermore, it is more accurate and has smaller variation than any other algorithm that relies on the distribution of signal strength in space. Gradient achieves a median AP localization error of 33 meters and reduces the maximum error by 33% compared to the second best, weighted centroid approach.

References

[1] Kim, M., Fielding, J.J., Kotz, D.: Risks of using AP locations discovered through war driving. In: Fishkin, K.P., Schiele, B., Nixon, P., Quigley, A. (eds.) PERVASIVE 2006. LNCS, vol. 3968, pp. 67–82. Springer, Heidelberg (2006)

[2] Cheng, Y., Chawathe, Y., LaMarca, A., Krumm, J.: Accuracy characterization for metropolitan-scale Wi-Fi localization. In: Proc. ACM MobiSys., Seattle, WA (June 2005)

[3] Subramanian, A., Deshpande, P., Gaojgao, J., Das, S.: Drive-by localization of roadside WiFi networks. In: Proc. IEEE INFOCOM, Phoenix, AZ (April 2008)

[4] Savvides, A., Han, C.C., Strivastava, M.B.: Dynamic fine-grained localization in ad-hoc networks of sensors. In: Proc. ACM Mobicom, Rome, Italy (July 2001)

[5] Rappaport, T.: Wireless Communications — Principles and Practice. Prentice-Hall, Englewood Cliffs (1996)

[6] WhitePages.com - White Pages Phone Directory, http://www.whitepages.com

[7] Bahl, P., Padmanabhan, V.: RADAR: An In-Building RF-based User Location and Tracking System. In: Proc. IEEE INFOCOM, Tel-Aviv, Israel (March 2000)

[8] Harter, A., Hopper, A., Steggles, P., Ward, A., Webster, P.: The anatomy of a context-aware application. In: Proc. ACM/IEEE MOBICOM, Seattle, WA (August 1999)

[9] Priyantha, N., Chakraborty, A., Balakrishnan, H.: The Cricket location-support system. In: Proc. ACM Mobicom, Boston, MA (August 2000)

[10] Fox, D., Hightower, J., Liao, L., Schulz, D., Borriello, G.: Bayesian filtering for location estimation. IEEE Pervasive Computing 02(3) (2003)

Management Tools

Extracting Network-Wide Correlated Changes from Longitudinal Configuration Data

Yu-Wei Eric Sung[1], Sanjay Rao[1], Subhabrata Sen[2], and Stephen Leggett[3]

[1] Purdue University
[2] AT&T Labs Research
[3] AT&T Inc.

Abstract. IP network operators face the challenge of making and managing router configuration changes to serve rapidly evolving user and organizational needs. Changes are expressed in low-level languages, and often impact multiple parts of a configuration file and multiple routers. These dependencies make configuration changes difficult for operators to reason about, detect problems in, and troubleshoot. In this paper, we present a methodology to extract network-wide correlations of changes. From longitudinal snapshots of low-level router configuration data, our methodology identifies syntactic configuration blocks that changed, applies data mining techniques to extract correlated changes, and highlights changes of interest via operator feedback. Employing our methodology, we analyze an 11-month archive of router configuration data from 5 different large-scale enterprise Virtual Private Networks (VPNs). Our study shows that our techniques effectively extract correlated configuration changes, within and across individual routers, and shed light on the prevalence and causes of system-wide and intertwined change operations. A deeper understanding of correlated changes has potential applications in the design of an auditing system that can help operators proactively detect errors during change management. To demonstrate this, we conduct an initial study analyzing the prevalence and causes of anomalies in system-wide changes.

1 Introduction

One of the most challenging tasks for IP network operators involves making and managing changes to router configurations that are needed to reflect changes in network designs, or as a response to address network problems. Configuration changes are often *system-wide* (involve most routers in a network) and *intertwined* (require modifications to multiple parts of a configuration file or localized groups of routers). Once configuration changes are made, these dependencies make it difficult for an operator to verify that the changes executed conform to his intent. Even worse, a small but incorrectly applied change can have serious impacts such as Service Level Agreement (SLA) violations for providers, and service disruptions for customer enterprises [1, 2, 3]. Yet, the goal of correctly making and effectively managing configuration changes remains daunting for operators, considering the large size and geographical span of networks, the myriad of configuration options, and the variety of routers from different vendors.

Existing tools (e.g., [4, 5, 6]) for automated change management are inadequate when coping with dependent changes for two reasons. First, typical tools are geared towards

S.B. Moon et al. (Eds.): PAM 2009, LNCS 5448, pp. 111–121, 2009.

managing one router at a time. Second, changes are tracked using device and vendor-specific low-level languages, and deal with myriads of details such as line card settings and routing parameters. Without a network-wide view of what changed and how changes were related, it is difficult for an operator to gauge the network state, verify changes were executed correctly, and know where to look for sources of potential or existing problems.

This paper introduces a methodology that extracts network-wide correlations of configuration changes (a group of changes that consistently occur together) and their high-level intent from low-level router configuration files. To do this, our methodology (i) identifies syntactic configuration blocks that changed by abstracting away low-level details, (ii) applies data mining techniques to expose correlated changes, and (iii) highlights changes of interest via operator feedback. We use router configuration files since they are considered by the operational community to be the most accurate source of records of changes. Distinct from prior works [7, 8, 9, 3] based on static configuration snapshots, we focus on developing longitudinal views of *changes across time*.

One distinguishing feature of our methodology is the use of data mining techniques. From our experience, operator knowledge tends to be incomplete. In particular, given the multitude of different configuration options, networks managed, and operator teams involved, it is nearly impossible for an operator to explicitly list all changes of interest up-front, not to mention that there may be hidden changes an operator may be unaware of. Employing data mining techniques enables automatic discovery of an initial set of correlated changes that are potentially important, without operator support. Yet, correlation does not always imply meaningful relationship. To address this, we corroborate the uncovered correlations with operators and only highlight meaningful ones.

Employing our methodology, we conduct a longitudinal study of changes made in enterprise VPNs, one of the most dynamic and demand-driven services that ISPs provide to customer enterprises today. We analyze a collection of daily snapshots of configurations files pulled from routers in operational networks. Our datasets include 5 enterprise VPNs, each consisting of a few hundred routers, over a 11-month period. Our analysis confirms the value and effectiveness of our methodology, and conveys important insights on change behavior in these networks. Finally, we conduct an initial study analyzing anomalies in system-wide changes as a demonstration of how our methodology can provide insights that help operators detect errors in change operations.

2   Methodology

2.1   Exploiting Syntactic Structure for Change Characterization

Existing router configuration languages are low-level and vendor-specific. The key issues we face are determining what the right abstraction is to associate a configuration change with and how to easily obtain the abstracted configuration for different vendor languages. We could abstract changes based on coarse semantic meanings (e.g., related to Wide Area Network versus Local Area Network), or low-level attributes (e.g., bandwidth 10 versus 100). However, both choices require detailed domain knowledge for the configuration language under study and are not feasible given the complexity and heterogeneity of today's configuration languages. We therefore choose to abstract changes based on the syntactic structure of configuration languages.

```
1 interface Ethernet0/0
2 ip address ......
3 !
4 interface ATM1/0
5 ip address ......
6 !
7 access-list 2 permit host 10.1.1.1
8 access-list 2 permit host 10.1.2.2
9 access-list 3 permit host 10.1.3.3
```

Fig. 1. A Cisco router configuration

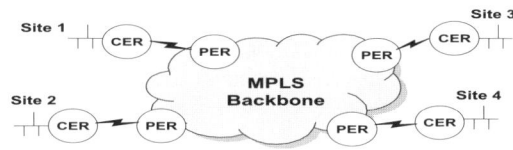

Fig. 2. An enterprise VPN with four sites

To demonstrate this in the Cisco IOS language, consider a Cisco router configuration file in Figure 1. The configuration may be viewed as consisting of multiple *blocks*. Each block comprises a first line that is not indented, and a series of indented lines associated with the block. For example, lines 1-3 constitute a block (interface Ethernet0/0). We also consider commands that lack a similar nested structure but share a common prefix, e.g., lines 7-8, to belong to a single block, in this case (access-list 2). For each non-indented line, we term the initial consecutive series of IOS keywords as a *superblock prefix*. Blocks with the same superblock prefix belong to the same *superblock*. For example, Lines 1-6 and 7-9 belong to superblock (interface) and (access-list), respectively. In this work, we focus on understanding configuration changes at the superblock level. Finally, our initial study of configuration languages from other vendors, such as Juniper and Alcatel, suggests that they bear similar block structures and are amenable to a similar approach.

2.2 Algorithm for Extracting Correlated Changes

To systematically extract correlated changes, we employ the *Apriori* algorithm [10], a powerful data mining technique for association rule induction. *Apriori* is typically used for *market basket analysis*, which aims at finding regularities in the shopping behavior of customers. It expresses an association between *items* within a *transaction*. An *association rule* is in the form "Given a transaction, if a set of items (or itemset) X occurs, itemset Y also occurs," or X→Y.

The standard measures to assess goodness of a rule are its *support* and *confidence*. Let T be the set of all transactions. The support of an itemset X, $S(X)$, is the percentage of transactions in T in which all items in X occur together. The confidence of a rule X→Y, $C(X→Y)$, is defined to be the percentage of times that the occurrence of X implies that all items in X and Y occur together, i.e., $S(X,Y)/S(X)*100\%$. Using these measures, an itemset X occurs frequently if $S(X)$ is high. A rule X→Y with a high confidence makes a good prediction about the occurrence of Y given that X occurs. If the average of $C(X→Y)$ and $C(Y→X)$ is high, itemsets X and Y are strongly predictive with respect to each other, i.e., they consistently occur together. We extend this idea to find *clustered* itemsets, where in each itemset R with n items, the average of C(Subset of $n-1$ items from $R→ n_{th}$ item) exceeds a threshold t_c.

We employ *Apriori* in the context of finding which routers or superblocks tend to change together in a network. For example, to find correlated routers, we define the set of transactions to contain days on which some router changed and the set of items in a transaction to be the routers that changed on a day. Similarly, to find correlated superblocks, we define a transaction to correspond to each instance of a router change on a given day, which we refer to ⟨router,day⟩, and the set of items to be the superblocks that changed in a router on that day. In our context, an association rule would be "If routers x and y change together in a day, router z always changes on that day," or "If some access-list (ACL) changes in a router, 50% of the time, some interface of that router changes, too."

3   Characterizing Changes in Enterprise VPNs

Fig. 2 shows a customer enterprise VPN spanning multiple sites over an Multi Protocol Label Switching (MPLS) provider backbone. Each site typically has a customer edge router (CER) connected via a WAN link to a provider edge router (PER). End-to-end Class of Service (CoS) is provisioned by marking packets and treating them differently according to their markings, on the CER-PER-backbone-PER-CER path. The dynamic and heterogeneous nature of changes to CERs [11] makes them the focus of our study.

Changes to CERs may be initiated by the customer and the provider. Events such as changes to passwords may be initiated by the provider as bulk updates in a VPN, or ISP-wide updates across multiple VPNs. Changes driven by customers might be planned, e.g., an interface or link upgrade. Unplanned changes might relate to troubleshooting customer complaints. A high-level change demand may involve a large number of CERs and operator teams.

Some changes are primarily controlled and maintained by the provider. This includes changes to passwords, packet and route filters, and management services like network time protocol (ntp) servers. Other changes, such as changes to CoS and routing designs, may be initiated by both the customer and the provider.

3.1   Datasets

Our data includes 11 months of daily archives of CER configuration files from 5 operational enterprise VPNs. We study longitudinal snapshots of configuration files for two reasons. First, configuration files are considered by network operators to be the ultimate and most accurate source of records of changes. Second, router configuration files are widely available in any network, ensuring our methodology is generally applicable. One data source we did not use is logs that show the sequence of low-level configuration commands executed by the operators. Such logs may enable us to directly reason about operator actions. However, information in these logs may be incomplete - it is possible for operators to bypass the logging system, particularly when bulk changes are involved.

Table 1 summarizes our datasets. All CERs are Cisco routers, and all 5 VPNs, E1-E5, are managed by the same provider. The total size of our data is 32GB. The networks were selected to cover a range of different characteristics in terms of size, geographic span, and growth. E3 and E5 had routers all in one country. At the other extreme, E2 and E4 spanned more than 30 countries and 5 continents. *Net Growth* and *Birth Rate* respectively represent the net change in network size (in terms of the number of CERs) and the total number of new CERs added over the 11-month period. Overall, E5 was the most stable due to its low Net Growth and Birth Rate. Configuration file sizes are also diverse within each VPN because different routers may have different roles (e.g., hub versus spoke), and different sites can have different local policies (e.g., a site hosting critical web services requires additional CoS and security configurations).

Table 1. Enterprise VPN data set. The number of CERs per network is between 150-420.

VPN	City	Ctry	Ctinent	Net Growth	Birth Rate	Config Size(# lines)		
						Min	Med	Max
E1	158	2	1	-1.47%	8.21%	408	1033	1487
E2	100	31	5	5.96%	16.56%	320	652	1175
E3	269	1	1	25.2%	25.2%	551	633	1622
E4	162	36	5	7.11%	25.26%	426	767	1475
E5	346	1	1	-2.85%	1.66%	436	489	1104

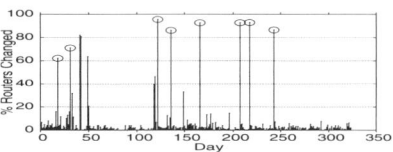

Fig. 3. Percentage of routers changed over time for E2

3.2 Macroscopic Overview of Configuration Changes

We now present key high-level characteristics of changes that we discovered across and within individual CERs.

● *Changes are predominantly local.* Across all networks, in 90% of the days, only 10% or fewer CERs were changed, and in 3% of the days the impact was widespread - with more than 80% of the CERs impacted. Fig. 3 shows the time series of the percentage of CERs changed for E2. Some system-wide changes (shown by high spikes) covered most of the CERs, and were often followed by changes to the remaining routers in the subsequent days. In addition, we found that many large-scale changes were correlated (circled spikes in Fig. 3) across networks. This is consistent with the fact that the provider may schedule ISP-wide changes to several VPNs in the same maintenance window.

● *Some routers are significantly more volatile.* For all 5 VPNs, almost all CERs (98.7%-100%) changed at least once, but the frequency was quite skewed. More than 90% of routers had changes on 3-6% of the days while a small fraction (around 2%) of routers were significantly more volatile, changing on 10%-35% of the days. We found that these volatile CERs usually corresponded to hub routers responsible for switching inter-spoke traffic. Therefore, changes to their configurations were often triggered by changes to spoke routers, e.g., adding an ISDN username/password for a new spoke site.

● *Most changes impact few superblocks.* Most configuration changes were limited to a small number of lines or localized to few superblocks. However, few changes impacted many lines or multiple parts of a configuration file. For example in E1, 58% of ⟨router,day⟩ instances had ≤10 lines changed while only 15% of them had >100 lines changed. In addition, 76% of ⟨router,day⟩ instances had changes to 2 or fewer superblocks.

● *Some superblocks consistently change more frequently.* We define the frequency of a superblock change as the percentage of the days that particular superblock changes in some CER per VPN. In all VPNs, superblock (interface) changed the most frequently while superblock (access-list) was among the top 10 frequent changes. Frequencies of changes to other superblocks were more varied. Other notably volatile superblocks were CoS-related: (policy-map) & (class-map), and routing-related: (router bgp).

4   Correlation Analysis of Changes

To demonstrate the value of our methodology in extracting correlated changes, we perform an in-depth analysis of the 5 VPN datasets. We then highlight particularly interesting correlated changes (i.e., system-wide and intertwined changes) that we corroborated with the operators managing these networks.

4.1   Correlated Changes across Routers in a Network

System-wide Changes. We consider days where a large fraction of routers in the enterprise changed. As shown in §3.2, system-wide changes could spread over a small number of days. Therefore, we consider a global event (or system-wide change) to be a window of w consecutive days where more than $f\%$ of all routers were modified. We pick $w = 2$ and $f = 80$ since Fig. 3 showed that most large-scale changes (high spikes) impacted $\geq 80\%$ of the CERs and were followed by few small-scale changes (short spikes). In the end, we identified a total of 51 global events across all 5 VPNs. This heuristic may miss events which impact all routers but are spread out over a prolonged period of time, and we discuss detecting some such changes in §4.2.

Next, to further understand the nature of global events, we apply *Apriori* (see §2.2) to extract superblocks that consistently changed together in CERs involved in each event. We observe that system-wide changes are typically homogeneous - in each global event, at least 80% of the CERs showed some change in one particular superblock. Among the 51 global events, only 8 events were related to CoS and the remaining were related to management and security operations. CoS-related changes were changes to ACL rules that specify flow memberships of traffic classes. All security changes were changes to ACLs and passwords to control access for remote telnet sessions and SNMP MIBs. Management changes were related to functions such as specifying when SNMP traps must be triggered, increasing the log buffer size, or setting the time zone.

Router Clusters. We consider correlations across small groups of routers that changed for each VPN. Note that a single global event impacts most routers and has the potential to skew this analysis. We therefore filter out days that were a part of some global event. We use *Apriori* to generate clustered router groups with t_c=80 for each VPN. Overall, 1, 4, 26, and 2 clustered router groups are reported for E1, E2, E4, and E5, with an average size of 2, 3, 9, and 3 CERs, respectively. A predominant trend is that the identified clusters show strong geographical proximity, with routers belonging to the same country, or continent. E2 and E4 have more clusters with a larger size on average because they are geographically widespread. Further discussions with the operators revealed a number of reasons for such regional clustering of changes. From a provider perspective, certain changes are administered by operators in different regions, while others are applied centrally. From the perspective of a customer enterprise, VPN sites in different regions may have different local needs, e.g., multiple hub-sites may be configured similarly in a primary-backup setup for resiliency reasons.

Table 2. Most frequently changed superblock groups within a router in E5

N-tuple	support (%)	conf.(%) of n-1 tuple
(access-list)	42.2	NA
(interface)	39.2	NA
(router bgp)	38.9	NA
(ip route)	38.9	NA
(access-list) (router bgp)	38.7	91.7-99.4
(access-list) (route-map)	38.6	91.3-100
(access-list) (ip route)	38.4	91.1-98.9
(access-list) (router bgp) (route-map)	38.2	98.9-100
(access-list) (ip route) (router bgp)	38.2	98.9-100
(access-list) (ip route) (router bgp) (route-map)	38.0	99.4-100
(username)	28.6	NA
(interface) (username)	28.4	72.5-99.6

Table 3. 3 most frequent superblock clusters for each VPN. C:CoS, M:Management, R:Routing.

Ent	Cat.	Superblock Group
E1	C	(interface) (policy-map) (class-map)
		(interface) (ip access-list extended) (policy-map) (class-map)
	R	(router bgp) (route-map) (ip access-list standard)
E2	C	(interface) (policy-map) (class-map)
		(interface) (access-list) (policy-map)
	M	(ntp server) (logging) (snmp-server host)
E3	C	(interface) (class-map) (policy-map)
	R	(interface) (access-list) (router ospf) (ip host)
	M	(logging) (ntp server) (snmp-server host)
E4	C	(interface) (policy-map) (class-map)
		(interface) (access-list) (policy-map) (class-map)
	M	(snmp-server host) (ntp server) (logging)
E5	C	(access-list) (interface) (policy-map)
	R	(access-list) (ip route) (router bgp) (route-map)
	M	(interface) (username)

4.2   Correlated Changes across Superblocks in a Router

Intertwined changes performed by operators may involve changing multiple superblocks. Our goal is to identify correlations across superblocks that consistently change together. To avoid skewing our results, we filter out days involved in global events.

Table 2 shows the groups of superblock(s) that change together most frequently for E5, sorted in decreasing order of support. Superblock (access-list) changes occur in 42.2% of the transactions (i.e., all ⟨router,day⟩ instances), while superblock (router bgp) changes occur in 38.9% of the transactions. Further, the pair of superblocks (access-list) and (router bgp) change together in 38.7% of the transactions. However, superblocks that change frequently individually need not change frequently together. For example, both (access-list) and (interface) individually change in over 40% of the transactions, but they change together in only 2.8% of the transactions (not shown). The right-most column summarizes the range of confidence values C(Subset of n-1 superblocks$\rightarrow n_{th}$ superblock) for all possible subsets of size n-1. For example, superblocks (access-list), (ip route), and (router bgp) occur together in 38.2% of the transactions, and if any two of the superblocks change, the percentage of times that the third superblock changes ranges from 98.9-100% depending on which two superblocks occur.

We now illustrate the types of correlated changes our methodology can identify.

Staggered System-wide Changes. A striking observation from Table 2 is that the group with superblocks (access-list), (ip route), (router bgp), and (route-map) occurs in 38% of the transactions. Further, for any 3-tuple combination of these superblocks, the confidence range is very high (>95%). Further investigation with the operator revealed that E5 experienced a change in its network design during the measurement period. The design moved away from using the provider's ISDN backup solution to a solution that points all traffic back to the customer environment in the event of the primary link failing, since the customer had added an additional service provider. The design change consists of modifications to the BGP configuration, additions of access-lists, route-maps and weighted static routes. These changes were introduced over a period of 2 months, configuring roughly 20-30 sites every 2-3 days. The system-wide change was spread over time to reduce the risk of adversely impacting the primary network traffic. Another superblock group, (interface) and (username), also has a relatively high support of 28.4, with a high confidence range. This turned out to be related to the second part of the same overall design change - removal of the existing ISDN backup solution, which involved deletions of usernames and logical ISDN interfaces, and modifications of physical interfaces referring to the removed logical interface. The slightly lower support for this group is because not all sites of E5 had an ISDN backup solution. Interestingly, these two groups of staggered design changes were performed by independent design teams, and the operators found our methodology useful in confirming these changes occurred as intended.

Frequently Occuring Superblock Clusters. Table 3 summarizes the 3 most frequent superblock clusters for each VPN. For enterprise E5, the 4-tuple corresponding to BGP policy addition, and the 2-tuple corresponding to ISDN backup removal are shown. For each superblock cluster, we assign a category of operation associated with the group. For all VPNs, we find that most intertwined superblock changes are centered around CoS (e.g., provisioning a new class of traffic) and routing (e.g., installing new backup routes), confirming the central role of CoS operations in all the VPNs we consider.

Syntactically Unrelated Meaningful Correlations. Table 3 shows that E2-E4 has a strong correlation in management operations related to ntp, logging and snmp-server. By merely looking at the configuration commands, it would not be clear how these superblocks are related since they do not directly refer to one another. Yet, they turn out to form a semantically meaningful correlation that reflects the periodic server update routines used in those VPNs. This type of correlation involves changes to syntactically unrelated parts of a configuration file. A parser incorporating knowledge of the configuration language itself would be incapable of extracting such correlation. This finding illustrates the potential benefits of our methodology in extracting syntactically unrelated, but semantically meaningful correlations.

5   Application – Finding Anomalies in System-Wide Changes

Knowledge of correlated changes has potential applications in detecting errors in the change management process. In this section, we conduct an initial study focusing on anomalies in system-wide changes.

A key observation we made when analyzing system-wide changes (§4.1) is that some system-wide changes impacted most, but not all, CERs. An auditing tool can leverage such insight and proactively look further for CERs missing an initial bulk update, which we call *outliers*. We analyze their prevalence and further investigate their causes based on operator responses.

We call outliers that never received the missed global update *persistent outliers*. Among the remaining outliers that eventually saw the missed global update, 80% of which received the update within 8 days. The operator indicated that these "short-lived" outliers' initial misses were due to network congestion or routers being overloaded, and were shortly fixed later by their auditing scripts. Therefore, we exclude them from our analysis and call the rest of outliers *delayed outliers*. Table 4 summarizes the outliers.

Table 4. Summary of global outliers detected. Numbers in parentheses denote the number of unique outlier routers and the number of events in which some indicated outlier occurred.

Ent	Total	Persistent Outliers			Delayed
		errors	non-errors	unknown	Outliers
E1	172(38,8)	0	134(26,8)	3(3,2)	35(12,6)
E2	24(15,6)	11(6,3)	7(7,1)	5(5,3)	1(2,1)
E3	9(8,2)	0	2(2,2)	7(7,1)	0
E4	91(85,7)	0	81(78,2)	6(6,2)	4(4,3)
E5	16(3,6)	0	10(10,1)	0	6(3,4)

Note that a CER may appear as an outlier multiple times if it missed more than one global update. We present some interesting causes for these outliers below.

• **Persistent Outliers:** We classified persistent outliers into *errors* and *non-errors*. In a few cases, we were not able to determine the causes, and we classified them as *unknown*. (i) *Errors*: These outliers were confirmed by the operators as needing fixes. We found 11 such outliers, all in E2. They corresponded to missed management updates, e.g., increasing size of logging buffers and setting timeout for management sessions. The operators indicated that although these errors were not critical to essential operations of the VPNs, it is important that all operators are aware of the existence of these errors in order to evaluate their potential impact, and take remedial actions if needed.

(ii) *Non-Errors*: These outliers were either confirmed or strongly suspected by the operators as genuinely not needing the update. They constitute the majority of outliers detected in E1, E4, and E5. The 134 cases in E1 involve only 26 routers, all related to CoS design. For example, a small fraction of CERs are located in a different country from all other CERs, and they use a different CoS design and have different update patterns. Non-errors in other networks were management-related. For example, low-end routers did not get the complete set of management ACL rules to reduce processing overhead. In addition, updates that increase certain parameter values (e.g., logging buffer size) above a threshold did not reach routers which already had them above the threshold.

• **Delayed Outliers:** Table 4 shows that E1 had the most delayed outliers, but on only 12 CERs. These routers used a newer Cisco style of CoS configurations which required manual updates because they were not amenable to bulk updates through older management tools. In addition, while E2-E5 allow fixes to be made on-demand, E1 had a more stringent update process in that changes can be made only in pre-scheduled time windows. These two factors explain a large number of delayed outliers in E1. For E2-E5, one major cause of delayed outliers was that the misconfiguration of management ACLs inadvertently blocked global updates.

6 Related Work

To our knowledge, the only other work that has analyzed dynamic operational tasks of real networks is [12]. While we share similar high-level objectives, [12] tries to identify groups of syntactically related commands and builds models to describe the series of actions needed for an operator to perform a given task on a router interface. In contrast, we focus on automatically extracting correlated changes within and across routers using data mining techniques. Minerals [7] also uses association rule mining to analyze configurations. However, they focus on detecting misconfiguration using static configuration snapshots.

Several works have sought to automate top-down generation of low-level configurations [11, 13] and typically focus on greenfield deployments. Others [3, 8, 9] have looked at detailed modeling and detection of errors in static configuration snapshots. Many of them are device-specific, and do not help operators understand and explain the semantics behind changes.

7 Summary

In this paper, we have presented a methodology to extract network-wide correlations of configuration changes from longitudinal snapshots of router configuration files. Our study of five operational enterprise VPNs over an 11-month period confirms the value and effectiveness of our methodology, and conveys important insights on the change behavior in these networks.

Our results show that while most changes affect individual routers, system-wide changes do occur, and primarily relate to management and security operations. In addition, correlations exist across groups of routers located in geographic proximity to each other. When correlations across superblocks were considered, most of these corresponded to changes to the CoS or routing design. Interestingly, one of the networks exhibited a markedly higher frequency of superblock groups - further analysis indicated this corresponded to a system-wide design change that was staggered over multiple days. Also of interest, our analysis revealed meaningful yet syntactically unrelated correlations, arising due to management processes employed in the networks.

While our findings are specific to the networks we analyzed, the methodology itself is generally applicable to all networks. A potential application is in tools that can provide operators with network-wide summaries of changes applied to their networks. Extracting correlations also has potential applications in the design of change auditing systems, that can alert the operator to violations of these correlations during a configuration change. We illustrate the potential of this direction by presenting an initial study of anomalies in system-wide changes in §5.

Much of this study has been performed with active involvement of operators, who have expressed great interest in the methodologies and findings. We are currently in the process of developing tools based on our methodologies for summarizing and auditing configuration changes.

References

[1] Mahajan, R., Wetherall, D., Anderson, T.: Understanding BGP misconfiguration. In: SIG-COMM (2002)
[2] Kerravala, Z.: Configuration management delivers business resiliency. The Yankee Group (2002)
[3] Narain, S.: Network configuration management via model finding. In: LISA (2005)
[4] Cisco IP solution center, http://www.cisco.com/en/US/products/sw/netmgtsw/ps4748/index.html

[5] Intelliden, http://www.intelliden.com/

[6] Opsware, http://www.opsware.com/

[7] Le, F., Lee, S., Wong, T., Kim, H.S., Newcomb, D.: Minerals: using data mining to detect router misconfigurations. In: MineNet (2006)

[8] Feamster, N., Balakrishnan, H.: Detecting BGP configuration faults with static analysis. In: NSDI (2005)

[9] Feldmann, A., Rexford, J.: IP network configuration for intradomain traffic engineering. IEEE Network Magazine (2001)

[10] Agrawal, R., Srikant, R.: Fast algorithms for mining association rules. In: VLDB (1994)

[11] Enck, W., McDaniel, P., Sen, S., Sebos, P., Spoerel, S., Greenberg, A., Rao, S., Aiello, W.: Configuration management at massive scale: System design and experience. In: USENIX (2007)

[12] Chen, X., Mao, Z.M., van der Merwe, K.: Towards automated network management: Network operations using dynamic views. In: INM (2007)

[13] Gottlieb, J., Greenberg, A., Rexford, J., Wang, J.: Automated provisioning of BGP customers. IEEE Network Magazine (2003)

Clarified Recorder and Analyzer for Visual Drill Down Network Analysis

Jani Kenttälä[1], Joachim Viide[1], Timo Ojala[2], Pekka Pietikäinen[3], Mikko Hiltunen[1],
Jyrki Huhta[1], Mikko Kenttälä[2], Ossi Salmi[2], and Toni Hakanen[2]

[1] Clarified Networks
Hallituskatu 9 A 21, 90100 Oulu, Finland
{jani,jviide,mikko,jyrki}@clarifiednetworks.com
[2] MediaTeam Oulu, University of Oulu
P.O. Box 4500, 90014 University of Oulu, Finland
{firstname.lastname}@ee.oulu.fi
[3] Oulu University Secure Programming Group, University of Oulu
P.O. Box 4500, 90014 University of Oulu, Finland
{pekka.pietikainen}@ee.oulu.fi

Abstract. This paper presents the Clarified system for passive network analysis. It is based on capturing complete packet history and abstracting it in form of different interactive high-level visual presentations. They allow for drilling from the high-level abstractions all the way down to individual packets and vice versa. The applicability of the system is demonstrated with the daily management of a large municipal wireless network.

Keywords: Iterative interactive network traffic visualization.

1 Introduction

We present the Clarified Recorder and Analyzer system for passive network analysis. It is a further developed and commercialized version of the freely available HowNet-Works tool that won the first price in VMware Ultimate Virtual Appliance Challenge in 2006 [5]. The system is based on capturing complete packet history, which is abstracted by various visual presentations. They represent different aspects of the packet data, for example individual events, identities, flows between identities, or causal relationships of flows. The visualizations facilitate visual drilling down from high-level visual abstractions to the level of individual packets and back. This facilitates high-level visual analysis of complicated network problems without tedious and time consuming detailed study of large amounts of packet data. Further, the availability of the captured packet data allows reactive assessment of security threats based on real traffic in the network. We describe the design and implementation of the proposed system and report its usage in the daily management of a large municipal wireless network.

2 System Design and Implementation

The architecture of the Clarified system is illustrated in Fig. 1. The Recorder captures traffic from one of more network interfaces creating a track, which is imported by the

S.B. Moon et al. (Eds.): PAM 2009, LNCS 5448, pp. 122–125, 2009.

Analyzer for iterative visual drill drown analysis. The Recorder uses an approach similar to the Time Machine [1], so that data collection agents (taps) are placed throughout the network. The Recorder collects packets from one or multiple taps, possibly filters the packet stream with a set of predefined filters, and stores the packets into multiple pcap files in a ring buffer fashion, together with a flow index computed on the fly [4]. The Recorder is able to simultaneously record and export the captured data to the Analyzer. The main design goals for the Recorder have been protocol independence and capturing of complete network traffic. The software components can be deployed on a standard off-the-shelf PC. The hardware dictates the scalability of the Recorder in terms of the effective packet rate that can be recorded without packet loss. The performance of the tap implemented in native C is limited by libpcap and the speed of the disk. With a high-performance RAID5 array and an optimized version of libpcap [6] multiple Gigabit Ethernet streams can be stored to disk without loss of data. The current bottleneck in the Recorder is the indexer written in Python, which is able to process about 40000 pps (~250 Mbps) on a high-end quad-core 2.6 GHz Xeon machine.

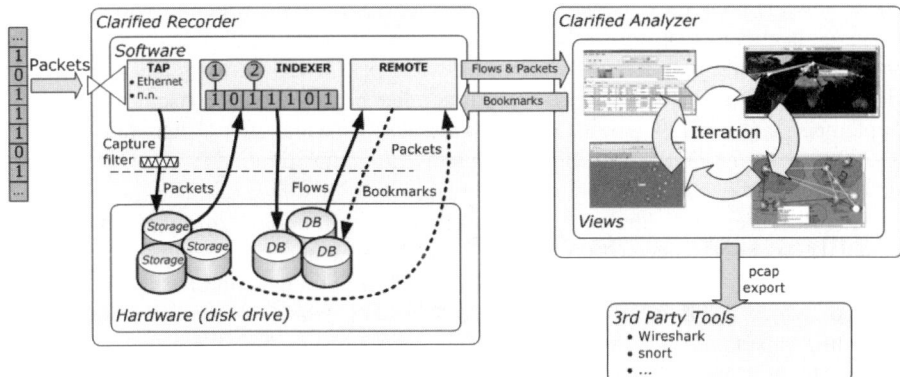

Fig. 1. The architecture of the Clarified system

The Analyzer imports a track (i.e. a flow index) created by a Recorder. The track is represented and manipulated on a timeline, which allows 'time travel' within the track. The Analyzer can conduct real-time analysis by importing a 'live' track that is currently being recorded by a Recorder. The Analyzer illustrates causal relationships [4] between different events in the track and other aspects with different types of visualizations, including *connection graph* (flows between identities), *layer graph* (relationships between entities), *association graph* (combination of connection graph and layer graph), *topology view* (flows rendered on a network diagram), *earth view* (data on a world map) and *DNS timeline* (TTL values of DNS resource records). The so-called monitors provide basic textual views of the packet data, including *flow monitor* (flows), *identity monitor* (identities) and *port monitor* (port statistics).

The main design goal for the Analyzer has been high-level analysis using 'living views'. It refers to the functional relationship between the visualizations provided by the Analyzer and the underlying packet data or network documentation. For example, given a particular visualization the user can access the corresponding packet data in real time, which allows for drilling down from a high-level overview down to the underlying flows and the details of individual packets if needed be. Similarly, visualizations can be mapped to network documentation so that the flows contained in the packet data are rendered on the network diagram, allowing visual inspection and understanding of the network traffic. Further, visualizations can be used to filter the input data, filtering affecting other views as well. For example, the earth view can be used for painting the geographical region of interest, on which further analysis will be limited. The packet data of interest can also be exported to 3rd party tools in pcap format. The AAA capabilities include centralized cookie-based user authentication, track-specific authorization, and accounting of Analyzer use and Recorder logins. The Analyzer is implemented in Python.

3   Usage in a Daily Management of a Municipal Wireless Network

panOULU is a municipal wireless network in the City of Oulu in northern Finland [3]. As of now the panOULU network has about 1050 WiFi APs, which provide open (no login, authentication or registration) and free (no payment) wireless internet access to the general public with a WiFi-equipped device. panOULU network has been operational since October 2003. The network usage has been growing constantly so that in September 2008 15127 unique WiFi devices used the panOULU network, totaling ~370000 sessions and ~13.9 million minutes of online time. For a comprehensive description of the panOULU network see [2].

The Recorder and Analyzer have been used in the daily management of the panOULU network since 2006. Three Recorder instances are deployed on a Dell PowerEdge 2900 with 4 CPUs and 8 GB of RAM. The machine has 3 TB of local disk, of which 2 TB is reserved for storing packet data. One Recorder instance is recording outside the NAT router, and two instances inside the NAT router. Capturing packets on both sides of the NAT allows inspecting possible NAT traversal problems, and deploying multiple Recorder instances with different filter configurations provides better scalability. Each instance is configured to record the first 100 bytes of each packet for the purpose of avoiding recording of private data possibly contained in the payloads. Since the average traffic is ~20000 pps, the 2 TB of disk is sufficient for recording ~11.5 days of traffic. So far peak rates have been ~71000 pps, which the multiple Recorder deployment has been able to capture without any packet loss.

The 'time travel' functionality is one of the main benefits of the Clarified system in the daily network management. If and when something abnormal happens, we can always easily return to the incident, as long as the data is still available on disk. A typical case is a rogue IPv6 router. Windows Vista operating system has an interesting 'feature': if a Vista machine is configured to use IPv6 routing (which is the default) and Internet connection sharing is enabled, the machine starts sending faulty IPv6 router advertisements, which mess up IPv6 routing in the network. As Vista is becoming more popular, so are rogue IPv6 routers in the panOULU network. To counter the problem we have implemented RogueIPv6Alerter as a Recorder helper application. Upon detecting a faulty IPv6 router advertisement it creates an alert bookmark identifying the event on the timeline. If we are able to locate the machine, we inform the user of the problem. If not, the machine is disconnected and placed on the list of blocked machines. The next time the machine tries to connect to the network it is automatically redirected to a web page explaining the problem and providing instructions for fixing the problem. Another typical case is an infected machine. We monitor the traffic for patterns involving large amounts of ARP traffic, which is typical for viruses. If and when such patterns are detected, we engage Analyzer to identify the infected machine for corrective action as with rogue IPv6 routers.

4 Conclusion

We presented the Clarified Recorder and Analyzer for visual drill down of network problems using different interactive visual abstractions of the captured packet data. By capturing complete packet history we guarantee that we can always drill down to individual packets when necessary. If we would sample the packet stream, there would be no chance to obtain the original flow information later. However, the trade-off is that the system does not scale up for high speed core networks. We are currently expanding the system towards wiki-based collaborative analysis.

Note. Videos illustrating the usage of the Analyzer are available online at http://www.clarifiednetworks.com/Videos.

References

1. Kornexl, S., Paxson, V., Dreger, H., Feldmann, A., Sommer, R.: Building A Time Machine for Efficient Recording and Retrieval of High-Volume Network Traffic. In: Internet Measurement Conference 2005, pp. 267–272 (2005)
2. Ojala, T., Hakanen, T., Salmi, O., Kenttälä, M., Tiensyrjä, J.: Supporting Session and AP Mobility in a Large Multi-provider Multi-vendor Municipal WiFi Network. In: Third International Conference on Access Networks (2008)
3. panOULU network, http://www.panoulu.net
4. Pietikäinen, P., Viide, J., Röning, J.: Exploiting Causality and Communication Patterns in Network Data Analysis. In: 16th IEEE Workshop on Local and Metropolitan Area Networks, pp. 114–119 (2008)
5. VMware Ultimate Virtual Appliance Challenge, http://www.vmware.com/company/news/releases/uvac_winners.html
6. Wood, P.: A libpcap Version which Supports MMAP Mode on Linux Kernels 2.[46].x., http://public.lanl.gov/cpw/

Data Gathering in Optical Networks with the TL1 Toolkit

Ronald van der Pol[1] and Andree Toonk[2]

[1] SARA Computing & Networking Services, Science Park 121, 1098 XG Amsterdam,
The Netherlands
rvdp@sara.nl
http://nrg.sara.nl/
[2] The University of British Columbia, 6356 Agricultural Road, Vancouver,
BC Canada V6T 1Z2
andree.toonk@ubc.ca

Abstract. This paper describes a new tool for gathering (performance) data from optical network devices. During the last few years many National Research and Education Networks (NRENs) have deployed hybrid networks. These hybrid networks consist of a traditional routed IP part and an innovative optical part. On the optical part *lightpath* services are offered. These lightpaths are high speed (up to 10 Gbps) circuits with a deterministic quality of service. Not all optical devices used in these networks offer full SNMP support, so the traditional management and monitoring tools cannot be used. Therefore we designed and implemented the TL1 Toolkit which supports management, data gathering and monitoring of optical devices.

1 Optical Networks and Lightpaths

In recent years many National Research and Educational Networks (NRENs) have acquired their own dark fiber infrastructure. The dark fibers are lit with DWDM and SONET/SDH equipment in order to build a circuit switched network that can be used for either transporting traditional IP traffic or so called lightpaths. Lightpaths are transparent point-to-point circuits with deterministic quality of service properties.

These optical networks provide researchers with new possibilities, but there are also significant research efforts in the infrastructure itself. The circuit switched nature of these optical networks is very different from the packet switched nature of the Internet. New technologies are being developed to monitor the health of lightpaths. Another topic of research is study which (performance) metrics are interesting to investigate in the optical networks. Section 3 discusses some of these new metrics.

2 TL1 Toolkit

The TL1 Toolkit is an open source Perl module that makes it easy to write Perl scripts that retrieve data from optical devices. Many optical devices use

S.B. Moon et al. (Eds.): PAM 2009, LNCS 5448, pp. 126–129, 2009.

the Transaction Language 1 (TL1) [1] as command line interface. Unfortunately, the syntax of TL1 is complex and the returned answers are difficult to read. Therefore, we designed and implemented a Perl module that hides the difficult syntax from the user and returns the results in easy to use Perl data structures. With the TL1 Toolkit it becomes easy to write Perl scripts that extract all kinds of data from optical network devices.

2.1   Supported Functions

This section describes some of the functions available. Below is a simple example that shows how to connect to a device, login, execute a TL1 command, logout and disconnect again.

$tl1 = TL1Toolkit->new("node", "name", "secret");
> Create a TL1Toolkit object that holds information about a TL1 session. Its arguments are:
> - hostname: network element to connect to
> - username: login name
> - password: password
> - type (optional): device type (OME6500, HDXc)
> - verbose (optional): debug level (0..9)

$tl1->open();
> Connect to the TL1 device and login. Returns 1 on success, 0 on failure.

my $output = $tl1->cmd($command);
> The string variable $command contains a TL1 command string that will be sent to the TL1 device as is. It returns the results as ASCII text to the string variable $output.

$tl1->close();
> Logout and disconnect from the TL1 device.

Another possibility is to use some of the build in functions, that will execute and parse the output of commonly used commands. A number of these build in functions are available. Examples are retrieving alarms, crossconnects, section trace data, inventory data and performance measurement data The example below retrieves the number of incoming octets on a specific interface.

my $octets = $tl1->get_inoctets(1, 3);
> Returns the Ethernet input octets on the slot given as first argument and the port given as second argument. Returns an integer.

3   Metrics in Optical Networks

The (performance) metrics in optical networks are different from the metrics used in the traditional routed Internet. Traffic in optical networks is transported over circuits instead of via packets. Parameters like RTT and jitter are not very interesting because of the deterministic nature of circuits. Instead new parameters such as the timeslot usage on links become important. Timeslot usage is an essential metric used in capacity planning for optical networks. Section 4 describes how this is done in the Dutch NREN (SURFnet6).

Other interesting metrics are the parameters involving the laser equipment, like voltages, power transmitted and received, optical loss, etc. This may give early warnings about degrading of the equipment or changes in optical impairments of the fibers. In general optical and circuits switched networks offer a wide array of performance measurement data. Retrieving this data and using it for analyses has proven to be very useful.

4   Lightpath Data Retrieval in SURFnet6

The TL1 Toolkit is used to retrieve data from the SURFnet6 network. SURFnet6 is a hybrid network and currently has about 200 lightpaths. Various configuration and performance data is retrieved from the SURFnet6 network. This is done by running TL1 Toolkit scripts periodically. The retrieved data is stored in a database. This data is made available via a web interface. The next section describes how this data is used for static lightpath planning in SURFnet6.

4.1   Lightpath Planning

The SURFnet6 lightpath part of the network consists of around 100 nodes and It soon became evident that with a network of this size manual planning of lightpaths was impossible. Therefore a planning tool [4] was written for SURFnet6. This uses the TL1 Toolkit in various ways. In order to find a path from A to B, the topology of SURFnet6 must be made available to the planning tool. The Network Description Framework (NDL) [5] of the University of Amsterdam is used to model the network topology. The NDL topology file is generated automatically by running a TL1 Toolkit script that discovers the topology information from the network utilizing the SDH section trace information.

Another script retrieves the used resource data of all the lightpaths from the network. This data is stored in the database. For each lightpaths the devices, interfaces and timeslots are stored.

The planning tool uses the NDL topology file to generate a mathematical graph of the network. In this graph a constraint based shortest path algorithm (Dijkstra) is used to find paths between two points in the network. The constraints are the timeslots already in use in the network and various links weights that represent network usage policies. Typically, the paths found are first reserved in the database so that no other lightpaths can use those resources. At a later stage, when all the formalities are completed, the lightpath is provisioned in the network.

5   Conclusions and Future Work

The TL1 Toolkit has proven to be a very powerful tool for writing Perl scripts with minimal effort to extract all kind of data from optical TL1 equipment.

Our own work with the TL1 Toolkit has been focused on monitoring operational status of lightpath [3]. But because there are many interesting metrics to investigate in optical networks, we will start tracking those too.

The introduction of dynamic lightpaths will introduce a whole new line of research. There is little operational experience with dynamic lightpaths and assigning timeslots in the most efficient way will be an interesting topic. The TL1 Toolkit is very well suited to retrieve and store data about dynamic lightpaths. This will produce a whole new set of network data that has not been available yet.

Acknowledgements. This work is supported by SURFnet and the GigaPort project of the Dutch government.

References

1. TR-NWT-000835, Telcordia Standard, Issue Number 03 (1993)
2. TL1 Toolkit download section, http://nrg.sara.nl/TL1-Toolkit
3. van der Pol, R., Toonk, A.: Lightpath Planning and Monitoring. In: eChallenges 2007 (2007)
4. van der Pol, R., Toonk, A.: Lightpath Planning and Monitoring in SURFnet6 and NetherLight. In: TNC 2007 (2007)
5. van der Ham, J., Grosso, P., van der Pol, R., Toonk, A., de Laat, C.: Using the Network Description Language in Optical Networks. In: 10th IFIP/IEEE International Symposium on Integrated Network Management, pp. 199–205 (2007)
6. GLIF Technical Report: Measuring the use of Lightpath service (2007), http://www.glif.is/working-groups/tech/lightpath-measurement-0.9.pdf

Audio and Video Traffic

A First Look at Media Conferencing Traffic in the Global Enterprise

Vijay Vasudevan[1,2], Sudipta Sengupta[2], and Jin Li[2]

[1] Carnegie Mellon University, Pittsburgh PA 15213, USA
[2] Microsoft Research, Redmond WA 98052, USA

Abstract. Many enterprise networks have grown far beyond a single large site to span tens to hundreds of branch offices across the globe, each connected over VPNs or leased lines. With the emergence of the globally-connected enterprise and the trend towards enterprise all-IP convergence, including transitioning the enterprise PBX to VoIP servers, IP-based audio/video conferencing for telepresence has challenged the notion that bandwidth is abundant in the enterprise. We take a first look at media conferencing traffic in the global enterprise and, by instrumenting measurement of call quality and network statistics, we quantify the impact on call quality for a range of factors in the enterprise, such as wired vs. wireless access, inter- vs. intra- branch office communication, QoS mechanisms like VLAN tagging and DiffServ DSCP marking, and VPN vs. public Internet access.

1 Introduction

Enterprise traffic analysis has received only scarce attention in the networking community, due in part to the difficulty of recording enterprise traffic. Early studies on enterprise traffic have found that network utilization is typically 1-3 orders less than network capacity, suggesting that enterprise networks have abundant bandwidth provisioning, unlike the WAN or global Internet [1]. This belief can no longer be assumed, primarily because of two trends: 1) the emergence of the globally-connected enterprise that spans countries and continents, with branch offices connected by leased lines, Virtual Private Networks (VPNs), and public Internet connections, and 2) the trend towards enterprise all-IP convergence, including transitioning the enterprise PSTN-based telephone exchange deployments to VoIP servers and IP-based audio/video conferencing for telepresence.

In particular, the move to VoIP and video conferencing is fundamentally changing enterprise traffic characteristics. Worldwide VoIP service is projected to grow to \$24.1B in 2007, up 52% from 2006. It is expected that worldwide VoIP service revenue will more than double over the next 4 years, reaching \$61.3B in 2011 [2]. To the best of our knowledge, there is no large-scale quantified measurement study of the quality of service for VoIP and video conferencing in the enterprise, despite its rapid expected growth.

In this paper, we attempt to characterize the network quality of audio/video conferencing and VoIP communication within a globally-connected and diverse

S.B. Moon et al. (Eds.): PAM 2009, LNCS 5448, pp. 133–142, 2009.

enterprise network. Using call logs from over 100,000 endpoints within this network over several months, we seek to understand several characteristics of enterprise network traffic and provisioning, including coarse statistics for packet loss, bandwidth utilization, and the causes of and potential solutions for poor call quality.

Our main results can be summarized as follows:

- Endpoints on wireless experience poor call quality as a result of a combination of non-negligible packet loss rates and high packet burstiness.
- Users connected to the enterprise over VPN or from home connections experience significantly higher packet loss rates.
- Non-wireless calls with DiffServ QoS priority treatment very rarely experience packet loss rates $> 2\%$, a loss rate which many audio codecs can mask.

We believe these results have implications for both enterprise network and protocol/application design, and suggest that large enterprise networks may not always be sufficiently provisioned given the growth of media conferencing traffic in the enterprise.

2   The Globally-Connected Modern Enterprise

Unified communication, which uses an IP-based network to carry many forms of communication traffic (e.g. VoIP, video conferencing, bulk data, and instant messaging) is gaining momentum in the industry. Most large enterprises today are turning their attention to unified communication to improve business processes, reduce communication delays, and improve employee productivity. While small to medium-sized enterprise networks consist of a small set of LANs within a single site [1], large, modern enterprise networks have grown far beyond this size, comprising several IP devices per employee, hundreds to thousands of employees per site, and tens to hundreds of sites around the world.

To accommodate this increasingly diverse geographic distribution, the scope of the modern enterprise has expanded to form a very heterogeneously-connected network. Using a large IT company with hundreds of branch offices across the globe as an example, our study focuses on the deployment and growing use of IP-based telephony and conferencing within this class of enterprise networks.

This network comprises the main campus networks located in the U.S., China, and Singapore via well-provisioned private lines (e.g. OC-48) and includes a large set of branch office sites, connected via VPN over the public Internet or over leased lines, with bandwidth capacity ranging from 1.5Mbps to 45Mbps depending on the size of the branch office. As a result, the modern large enterprise differs significantly from typical single-site enterprise deployments. Even within a single site, devices will use wireless or wired connectivity. Although the network is managed by one administrative body, the distribution of branch offices makes it difficult to provision all offices systematically to meet (variable) traffic demands. As a result, the assumption that enterprise networks are likely to have sufficient bandwidth throughout no longer holds.

As a step towards understanding such large-scale enterprise networks, in this paper we try to characterize one aspect of traffic within the network: IP telephony and audio/video conferencing, a growing portion of traffic that requires strict guarantees on performance for effective communication. While still preliminary in scope, we hope this investigation provides another initial step towards understanding the use and deployment of enterprise networks and motivates further study in the area.

3   Measurement Methodology and Datasets

3.1   Measurement

Our datasets consist of several IP phone and audio/video conference call log databases. A call can span multiple users at multiple branch offices. Each branch office has one or more *Media Control Unit (MCU)* servers whose function is to bridge the call – an audio/video stream originating at a user endpoint (client) is received at an MCU and replicated to all other participants (the choice of the MCU location for a given call depends on the location of the earliest joining user in the call). Thus, a call consists of multiple *sessions*, each session identified by a {user endpoint, MCU} pair. For a two-party call, the MCU is not involved, hence the session identified by a {user endpoint, user endpoint} pair. If any of the conference participants is on the PSTN (Public Switched Telephone Network), then a PSTN-gateway server is also involved.

For every IP phone call or audio/video conference call made within the enterprise, a *session-level* report is logged at the end of the call by participating endpoints to a central reporting server. Each individual session log contains several pieces of information for streams in both directions, including start and end timestamps, average/max packet loss, latency, average/max jitter, bandwidth capacity estimate, network quality (also known as *Network Mean Opinion*

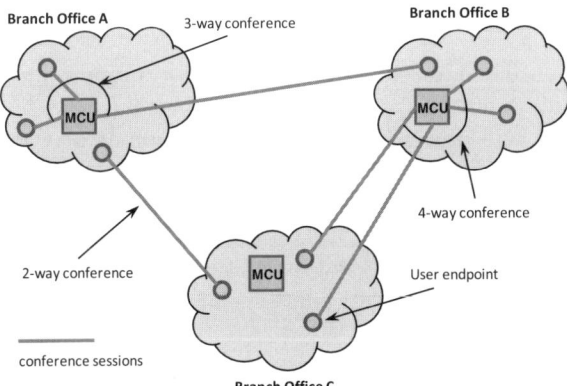

Fig. 1. Illustrating sessions in multi-party audio/video conference calls across branch offices – the session is the level at which measurements are logged

Score, or *NMOS*), device type, audio/video codec used, bitrate, etc. While the dataset does not include packet-level traces, these call log summary statistics are useful for identifying the cause of poor call quality in the enterprise, and allow for a first-look at new kinds of modern enterprise traffic characteristics. Although packet-level traces would allow for detailed traffic analysis, we instead focus on a broad characterization of call traffic in the enterprise by analyzing call volume, the effect of QoS, the real-world impact of wireless on call quality, and individual call statistics to identify the growth and potential implications for future enterprise network design.

3.2   Datasets

Table 1 describes two datasets we obtained for our characterization study. Dataset D_1 consists of a log set for a "dogfood" test deployment within the enterprise, while D_2 contains a recent, full log database of all calls. Because our logs comprise usage data for over 100,000 endpoints, our study focuses on a broad characterization of the pervasive IP-based conferencing traffic statistics in the enterprise network. We note that the number of distinct endpoints is larger than the number of users in the enterprise, because many users make calls from several endpoints (e.g., desktop machine, wireless laptop, IP phone). We use the D_2 dataset for all our analysis due to its increased log density and recency, but mention D_1 here only to note that the results obtained from both datasets are similar (except for call density), suggesting that the results presented in this paper cannot simply be attributed to initial deployment and provisioning hurdles.

Table 1. Overall Dataset Characteristics

Data	Start Date	End Date	#Audio Streams	#Video Streams	Distinct IP Endpoints
D_1	09/12/07	01/16/08	532,191	49,235	17,118
D_2	01/21/08	06/22/08	9,744,660	617,018	205,526

We supplement our datasets with location information pulled from an internal database that contains the location of all IP subnets within the enterprise network. More specifically, the location information is of the form {IP Subnet/mask, Country, City, Building}. For each IP address in our dataset, we perform a longest-prefix match to map the location of the caller to a specific building, city, and country. For security and privacy reasons, however, we do not obtain router-level topology information for the enterprise network. In total, the location database contains entries for over 7,000 subnets spanning 500 unique buildings/locations around the world.

4   Data Analysis

In analyzing our datasets, we break down the results into two categories: trends and statistics, and quality diagnosis. The first category discusses overall media traffic characteristics within the enterprise, such as aggregate call patterns, and overall traffic trends. The second category concerns evaluation of call quality and diagnosis of aspects that can impact call quality, such as QoS mechanisms like DiffServ [3], wired vs. (last-hop) wireless usage, and integration with the public Internet.

4.1   Trends and Statistics

Growth Trends: In Figure 2, we plot the number of audio and video streams per day over a six month period from D_3. During this period, total call volume nearly tripled as a result of an aggressive IP phone deployment across the enterprise starting in April 2008. On the other hand, the number of video streams did not increase as dramatically over the six month span, showing 60% growth over six months. We note, however, that the more gradual video stream growth is not connected to any particular external deployment effort and represents a trend towards increased video conferencing.

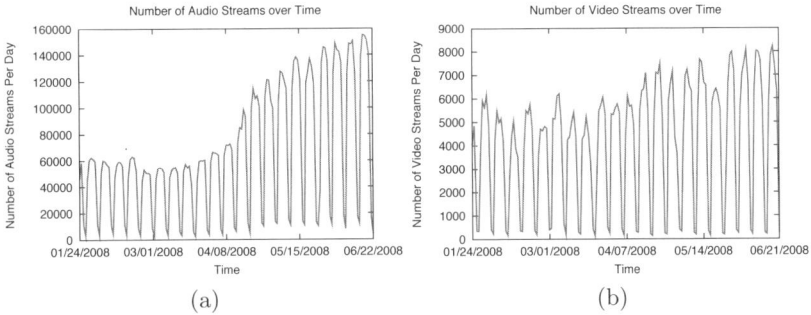

Fig. 2. (a) Audio trends follow call trends closely; growth in call volume was mostly due to increased VoIP and audio conferencing traffic. (b) Video calls increase slower in frequency in comparison to audio calls.

Call Patterns: We note several types of patterns in our dataset, ranging from time-of-day and time-of-week effects, to conference call patterns. Figure 3(a) shows the typical media traffic patterns observed at intermediate media devices (PSTN-gateways or conference stream aggregators) over a representative week. As expected, traffic exhibits a strong time-of-week effect, with peak usage during mid-week and a strong drop-off on weekends. Figure 3(b) depicts the same data for a typical Wednesday, with traffic always seen between business hours and occasional peaks due to individual large conferences. We also note the existence of many larger conference calls that occur at weekly intervals, predictably producing media traffic peaks.

4.2   Call Quality Diagnosis

Feedback indicating poor quality calls motivated us to investigate the cause of such calls and the impact of some efforts to improve call quality. Several factors

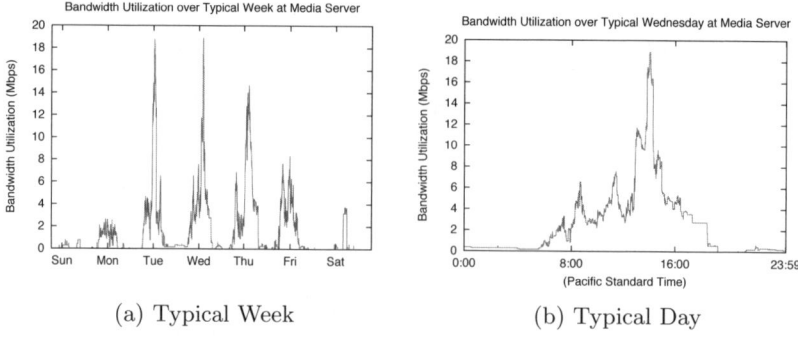

(a) Typical Week   (b) Typical Day

Fig. 3. Call traffic seen at a typical media aggregator

lead to poor call quality, including network-specific factors like packet loss, jitter, delay, but also application-specific factors such as audio echo, codec bitrate and frame loss rates. While Mean Opinion Scores (MOS) do help differentiate poor calls from acceptable ones, we find that max packet loss is one important contributor to poor calls from the network standpoint. In this dataset, packet loss is calculated over time intervals of 30 seconds during the call, with the maximum packet loss interval logged at the end of the call. Max packet loss is a good indicator of a poor call because even a single moment of high packet loss can distract callers and lead to a poor opinion of call quality. To this end, we focus on identifying the cause of, and potential remedies for high packet loss.

Wireless Last-Hop: While enterprise networks are often comprised of desktop and server machines on wired LANs, a significant number of hosts in the large enterprise we study actively use wireless LANs, a trend that is likely to increase given advancements in wireless deployments and ease of use. Much research has focused on the interplay between VoIP and 802.11-based wireless access, noting poor VoIP call quality on these wireless deployments despite having sufficient bandwidth capacity [4, 5]. In Figure 4(a), we plot the CDF of max packet loss, distinguishing between streams with endpoints communicating over wireless last hops. Streams involving wireless experience much worse packet loss characteristics than those on wired connections. For example, 5-10% of streams experience packet loss rates above 4%, producing poor audio quality and yielding a very poor user experience for those using video conferencing on these wireless devices. Although calls where both endpoints are wireless tend to be worse than with just one wireless endpoint, the marked difference is between wireless and non-wireless calls.

Coarse packet loss statistics alone do not tell the entire story. Most telephony and conferencing protocols are also sensitive to the duration of packet loss bursts, the length of time for which a large fraction of packets are discarded because of delayed arrival. As shown in Figure 4(b), audio streams with at least one wireless endpoint experience more and longer burst durations on average than wired connections. Thus, even if the wireless streams do not drop packets, the packets that do arrive may be useless to the real-time application.

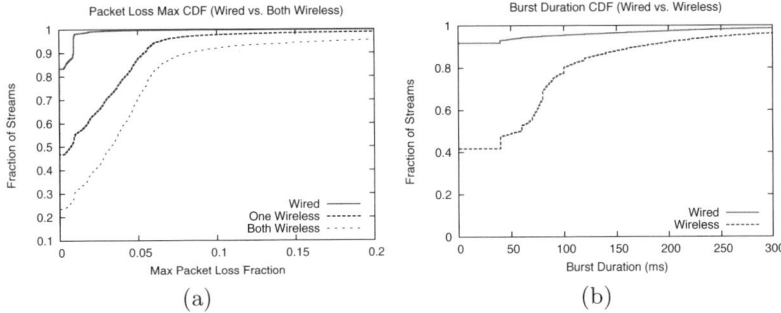

Fig. 4. CDF of packet loss fraction and burst duration broken down based on whether endpoints have a wireless last hop

Given the prevalence of (IEEE 802.11-based) wireless users (e.g., those with laptops in conference rooms), these real-world results motivate the need for deployment of newer VoIP and conferencing protocols that can accommodate such high-loss rates and minimize burst without a large degradation in call quality.

Home Users and VPN Clients: While calls within the enterprise experience higher than expected packet loss rates, we compare intra-enterprise calls to those calls that transit part of the public Internet to better understand whether the issue is specific to this enterprise network. Figure 5 plots the CDF of packet loss, separating in-enterprise endpoints and VPN/home endpoints. While packet loss within the enterprise is higher than expected, packet loss for VPN and home users is very high: 5% of streams experience packet loss rates greater than 10%. This suggests that the enterprise network is indeed better provisioned than the public Internet, though may not be as underutilized as previously noted [1]. We also observe that for external users, the call quality is perceivably better over the public Internet than over VPN. For example, an additional 10% of VPN streams experience packet loss of 5% or higher than public Internet streams. This can be attributed to VPN servers not adequately handling voice traffic, since they were designed with email/corporate-intranet traffic in mind.

Without packet-level traces or other data-traffic originating from or terminating at those endpoints, it is difficult to conjecture why these rates are much higher and more prevalent across the public Internet. For example, these higher loss rates may be caused by network congestion at the endpoint's ISP, wireless packet loss at the home user's network, or local congestion with other bulk data traffic from the endpoint itself.

QoS and Voice VLAN Usage: Most IP phone deployments are configured to exist on a separate voice VLAN [6]. Packets sent from the IP phone are marked with prioritized DiffServ DSCP [3] bits. Core routers within the enterprise and outgoing inter-branch office interfaces prioritize this voice VLAN traffic over normal best-effort data traffic. In Figure 6(a), we plot the CDF of the max packet loss fraction broken down based on whether zero, one, or both endpoints exist on

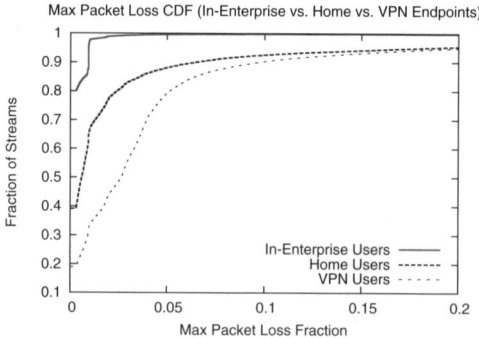

Fig. 5. Maximum packet loss CDF depending on type of user: within enterprise, external over public Internet (home), and external over VPN

the voice VLAN. In this graph, we remove calls that involve wireless endpoints to focus on the benefits of QoS on the wired LAN. Further, we remove endpoints outside the enterprise network (e.g., calls from employee homes, through VPNs) to avoid paths that traverse the public Internet, which does not respect DiffServ priority.

When both endpoints are on the voice VLAN, nearly all calls experience less than 2% packet loss, a loss rate that most audio codecs can accommodate. When exactly one endpoint is on the voice VLAN, max packet loss increases: 25% of audio streams exhibit some non-negligible packet loss. However, most calls do not experience more than 4% packet loss. When neither endpoint is on the voice VLAN, we find that 2% of streams experience packet loss rates greater than 4%. These results suggest that QoS for voice traffic can mostly eliminate the prevalence of poor audio calls due to packet loss in the wired enterprise.

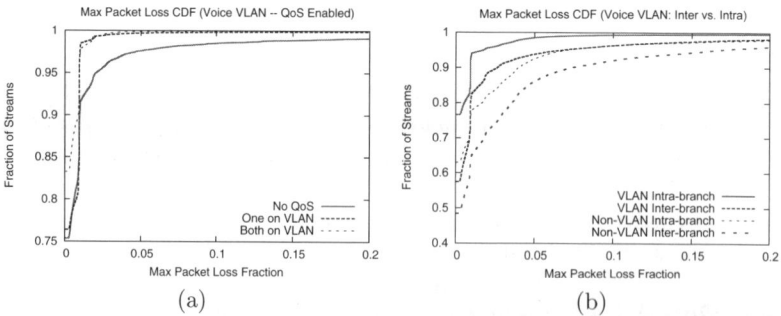

Fig. 6. (a) CDF of max packet loss fraction broken down into endpoint QoS support. Calls that travel on the voice VLAN are DiffServ-enabled and prioritized by core routers and at inter-branch routers. (b) Max packet loss broken down based on whether call involves VLAN and whether call traverses inter-branch office links.

Because QoS is more likely to help on congested paths, we further break down packet loss characteristics based on whether the call traverses an inter-office branch link or remains within the same branch office in Figure 6(b). Given the same QoS capability, packet loss rates on intra-branch paths are lower than on inter-branch paths, suggesting that inter-branch paths are likely bottlenecked at a common chokepoint. Interestingly, inter-branch office calls with QoS enabled tend to perform better than intra-branch office calls without QoS: there is enough congestion within a single site that intra-branch calls without QoS can suffer more than prioritized inter-branch calls. Thus, even the assumption that single sites are well-provisioned within an enterprise network may not be true.

5   Related Work

Measurement studies of enterprise networks have, up until recently, been scarce, despite the growth of the enterprise sector within industry. In the past decade, Shaikh et al. studied OSPF behavior in an enterprise network [7], and Pang et al. provided traffic analysis of a modern, medium-sized enterprise network, focusing on categorizing the types of traffic and network behavior seen in the enterprise based on packet traces [1]. Recent studies have also focused on enterprise network management [8], troubleshooting [9], or wireless diagnosis [10]. More recently, Guha et. al identify mobile (often off-site) hosts for degraded "health" in enterprise networks, where health is defined as the fraction of useful flows [11].

Our paper looks away from operational aspects of enterprise networks and instead attempts to shed light on an important and growing class of traffic within the modern enterprise: VoIP and conferencing. In addition, we provide a first look into a new class of modern enterprise networks that spans the globe and is heterogeneous in connectivity, making such networks more like the wide-area Internet in scope.

6   Conclusion

To the best of our knowledge, this is the first work in the literature to measure and analyze media traffic in the global enterprise. Using session-level reports logged at the end of each call containing call quality and network statistics, we quantify the impact on call quality for a range of factors in the enterprise, including wired vs. wireless access, inter- vs. intra- branch office communication, QoS mechanisms like VLAN tagging and DiffServ DSCP marking, and VPN vs. public Internet access. With the modern enterprise spanning the globe, the transitioning to VoIP from PBX, and rich media applications like voice and video conferencing seeing explosive growth in the enterprise, it can no longer be assumed that bandwidth is "abundant" in the enterprise. Our studies indicate that careful network planning and provisioning may be required in the enterprise to support good quality for media applications and that QoS mechanisms for prioritized traffic handling do indeed help to improve quality. Our continuing work in this area involves further analysis of media traffic in the enterprise using packet level traces and active measurement.

References

[1] Pang, R., Allman, M., Bennett, M., Lee, J., Paxson, V., Tierney, B.: A First Look at Modern Enterprise Traffic. In: Proc. ACM SIGCOMM Internet Measurement Conference (October 2005)

[2] Machowinski, M.: VoIP services and subscribers: annual market share and forecasts. In: Mobile Computing and Communcations Review (2008)

[3] Blake, S., Black, D., Carlson, M., Davies, E., Wang, Z., Weiss, W.: An Architecture for Differentiated Services. Internet Engineering Task Force, RFC 2475 (December 1998)

[4] Wang, W., Liew, S.C., Li, V.: Solutions to Performance Problems in VoIP over 802.11 Wireless LAN. IEEE Transactions on Vehicular Technology (January 2005)

[5] Niculescu, D., Ganguly, S., Kim, K., Izmailov, R.: Performance of VoIP in a 802.11 Wireless Mesh Network. In: Proc. IEEE INFOCOM (March 2006)

[6] IEEE 802.1Q Virtual LAN Standard,
http://www.ieee802.org/1/pages/802.1Q.html

[7] Shaikh, A., Isett, C., Greenberg, A., Roughan, M., Gottlieb, J.: A Case Study of OSPF Behavior in a Large Enterprise Network. In: Proc. ACM SIGCOMM Internet Measurement Workshop (November 2002)

[8] Casado, M., Freedman, M.J., Pettit, J., Luo, J., McKeown, N., Shenker, S.: Ethane: Taking Control of the Enterprise. In: ACM SIGCOMM 2007 (August 2007)

[9] Bahl, P., Chandra, R., Greenberg, A., Kandula, S., Maltz, D.A., Zhang, M.: Towards Highly Reliable Enterprise Network Services Via Inference of Multi-level Dependencies. In: ACM SIGCOMM 2007 (August 2007)

[10] Cheng, Y.C., Afanasyev, M., Verkaik, P., Benko, P., Chiang, J., Snoeren, A.C., Savage, S., Voelker, G.M.: Automating Cross-Layer Diagnosis of Enterprise Wireless Networks. In: ACM Sigcomm 2007 (August 2007)

[11] Guha, S., Chandrashekar, J., Taft, N., Papagiannaki, K.: How Healthy are Today's Enterprise Networks? In: IMC 2008: Proceedings of the 8th ACM SIGCOMM conference on Internet measurement (2008)

Supporting Enterprise-Grade Audio Conferencing on the Internet

Krishna Ramachandran and Sunitha Beeram

Citrix Online
Santa Barbara, USA

Abstract. This paper evaluates if the Internet can support enterprise-grade audio conferencing. For our investigation, we collect real-world traffic traces from the audio conferencing solution in GoToMeeting, a popular online meeting application. We analyze these traces using *Conference MOS*, a new metric proposed in this paper to measure the quality of an audio conference. Our results indicate that a majority of users experience good conference quality. A small percentage of users that experience poor quality because of high delay and loss would be better served with PSTN. This leads us to believe that an enterprise-grade solution should adopt a combined VoIP and PSTN deployment strategy over a VoIP only solution.

1 Introduction

The market for audio conferencing and its integration with enterprise applications is growing fast. For example, in Europe, it is projected to reach USD 712 Million in 2010 [3], while in the United States it is expected to be up to USD 3 Billion by 2013 [5].

Audio conferencing has traditionally been available via the Public Switched Telephone Network (PSTN). Enterprise users use their PSTN phones to dial into meetings hosted on conferencing bridges. With the rapid evolution of VoIP technologies and the increasing penetration of broadband access in residential and business settings, audio conferencing over VoIP is fast becoming a viable alternative.

Models for VoIP conferencing include *dial-in conferencing* and *peer-to-peer conferencing* [18]. In dial-in conferencing, end-points call into a Voice Conferencing Bridge, which performs audio mixing and conference control. In the peer-to-peer model, the end-points are responsible for these tasks. An example of the latter approach is Skype [6], which by most measures epitomizes VoIP telephony on the Internet.

Judging by the popularity of Skype, one can hypothesize that peer-to-peer mixing is well suited for conferencing. However, supporting *enterprise-grade* quality presents a set of critical challenges, not seen in typical peer-to-peer applications. We define enterprise-grade conferencing as one that is scalable, available and has Mean Opinion Score (MOS) [9] quality rating higher than 3.5.

In terms of scalability, enterprises require that audio conferencing solutions support a large number of callers, sometimes up to a thousand users within a single conference [2]. The processing and packetization of audio are complex operations. For example, Skype, which operates by mixing audio at the end-points, can only support a

S.B. Moon et al. (Eds.): PAM 2009, LNCS 5448, pp. 143–152, 2009.

conference up to 25 users [6], because of this complexity. In order to support a large number of users, powerful servers are required to perform audio processing and encoding. Dial-in conferencing is amenable to such provisioning.

The second challenge has to do with availability. Enterprises expect high availability, 99.999% being the holy-grail. Users in a *large* meeting can join and leave whenever they choose. In a peer-to-peer setting, handling user churn and failures is challenging, both in design and implementation. In contrast, dial-in conferencing is simpler to design and implement, operate, troubleshoot in case of problems, and, finally, provision for fault-tolerance and reliability.

In this paper, we are interested in investigating how well an enterprise-grade audio conferencing solution performs on the Internet. Toward this goal, we characterize the performance of our dial-in conferencing solution that provides audio conferencing for Citrix GoToMeeting [1], a popular online meeting application used by thousands of businesses around the world. GoToMeeting is a Sofware as a Service (SaaS) application that is used to collaborate online, give presentations and for online training. For our study, we collect five days of traffic traces from Aug 5-9, 2008 from a subset of voice conferencing bridges that are part of our VoIP infrastructure. Our data set contains over 9000 VoIP flows, which total approximately 230000 minutes of talk time.

For our analysis, we propose the Conference Mean Opinion Score (CMOS), a new metric that helps characterize audio quality in a conference. The MOS of a conference depends not only on the number of users in a conference and their network connections, but also on the time varying set of active speakers. Existing metrics fail to account for these considerations.

The major findings from our study are as follows:

– A majority of users experienced good quality in their conferences. The median CMOS was approximately 4.27. Only 11.75% of users had CMOS less than 3.5.
– A majority of VoIP connections to the conferencing bridges experienced good network conditions. The median MOS of these connections is 4.30. Only 11% had MOS less than 3.5.
– Unlike VoIP paths in the backbone that are well provisioned for high capacity [12,14], we find that end-to-end paths can experience high delay and loss. These factors resulted in 11% of VoIP connections to have MOS less than 3.5.

This paper offers the first measurements-based characterization of VoIP conferencing on the Internet based on real-world traffic measurements. Past VoIP studies [12,13,14,16] primarily investigated the performance of peer-to-peer VoIP flows. A second contribution of this paper is the Conference MOS, a metric for the estimation of audio quality in a conference.

2   Related Work

Several studies [12,13,14,16] have investigated the performance of VoIP flows in sections of the Internet. Bolot [13] in 1995 found that loss was a result of network congestion. Markopoulou et. al. [16] in 2001 probed 43 backbone paths between five cities in the US. They found that some of these paths experienced delay spikes and intermittent outages lasting 0.5-2 seconds. In a more recent study, Birke et. al. [12] monitored the backbone of an Italian ISP and found loss, not delay, to be the principal contributor to bad quality. Majority of VoIP calls placed over the ISP backbone were of good quality. Boutremans et. al. [14] reported that loss is a result of failures in backbone links because of either inefficiencies in the routing protocol, faulty router software or human errors.

To the best of our knowledge, there exists no measurements-based characterization of a real-world audio conferencing solution. Birke et. al. [12] studied VoIP calls placed in the backbone network of an Italian ISP. However, they did not focus on audio conferencing and did not consider end-to-end VoIP flows that span multiple domains.

Proposals [8,17] exist to estimate the Mean Opinion Score (MOS) [9] of a conversation between two participants. To the best of our knowledge, there is no solution for the estimation of the MOS in a conference. Conference MOS depends on the number of speakers, the quality of their network connections and the active speaker set at any given time. The solution offered in this paper takes these factors into account.

3   Conference MOS Estimation

We base our estimation of the conference MOS on the *E-model* [8], a widely used analytic model standardized by the International Telecommunications Union (ITU). The E-model predicts MOS by taking into account various factors, such as equipment quality, encoder used, packet delay and loss. We use the E-model because it has two main advantages: (1) it is *non-intrusive*, i.e., it does not require comparison of the original voice signal against a degraded one; and (2) it is *parametric*, i.e., it predicts MOS from measured properties, such as echo, delay and loss. In contrast, *signal-based* models [17] function by processing the output voice signal. Parameteric models, in general, are easier to apply than signal-based methods. Further, where user privacy is a critical requirement, such as in enterprise solutions, the use of a signal-based method is typically not possible.

3.1   E-model Overview

The basic principle behind the E-model is that the perceived effects of impairments are additive. The E-model results in a quality rating R on a scale of 0 in the worst case to 100 in the best case for narrow-band codecs and from 0 to 129 for a mix of narrow-band and wide-band codecs[10]. Figure 1 shows the relation between MOS and R for wide-band codecs.

R is computed as: $R = 129 - I_d - I_e$, where 129 is used, because our data set has flows that are encoded using *iSAC* [4], a wide-band codec; I_d and I_e are impairment factors that represent the quality degradation due to delay and loss respectively. Notice that the two impairments are isolated into separate terms even though the delay and loss events may be correlated in the network. This is because the E-model computation assumes that the impairment contributions corresponding to the network events are separable.

In actuality, R takes into account several other impairments, such as echo, background noise and imperfections in end-user equipment. These are not represented above

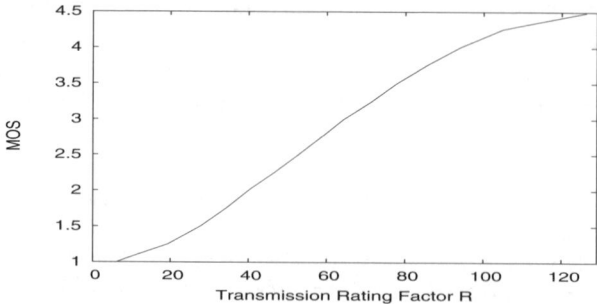

Fig. 1. Mapping between R and MOS for wide-band codecs

because we are primarily interested in assessing the suitability of the Internet for audio conferencing, whereas these factors do not depend on the network [15].

Impairment values for the iSAC codec are not available in ITU standards. Hence, we derived these values empirically as described in ITU specifications [7]. We conducted conversation-opinion tests [7] by varying either the network delay or loss. We used ten pairs of human subjects in our testing. For delay, the two test subjects took turns counting numbers as fast as possible; delay effects are most noticable during such a test. For loss, each subject read ten lines from a random paragraph. Each reading lasted about thirty seconds which is sufficient time for a subject to observe any degradations [17]. Figure 2 plots the average of the impairment values obtained from the ten experiments.

3.2   Conference MOS

Existing metrics fail to account for two important considerations when assessing the quality of a meeting hosted on a Voice Conferencing Bridge (VCB): (1) the voice quality perceived by a participant not only depends on the quality of its path to the VCB, but also on the number of *speakers* and the quality of the network paths between the speakers and the VCB; and (2) the set of speakers is likely to change on a frequent basis during the course of the meeting; therefore, quality measurements should periodically account for the up-to-date speaker set.

The following metric accounts for the above considerations:

$$cmos_p^t = \frac{\sum_{i=0}^{m} mos_{ip}}{m}$$

where $cmos_p^t$ is the *short-term* conference MOS for participant p for time period t, mos_{ip} is the MOS, computed using the E-model, on the path between speaker i and participant p traversing through the VCB v, and m is the number of speakers where $p \notin m$ if p is a speaker. For the computation of mos_{ip}, the path delay is given as $D = d_{iv} + d_{vp}$, where d_{iv} and d_{vp} are delays between i to v and v to p respectively. The path loss is given as $L = 1 - (1 - l_{iv}) * (1 - l_{vp})$ where l_{iv} and l_{vp} are the respective loss probabilities.

The parameter t controls the number of speakers that are included in the quality estimation. In the above equation, equal weight is given to each speaker in the time period t. If t is large, such weighting can skew the result. For example, if t is a minute,

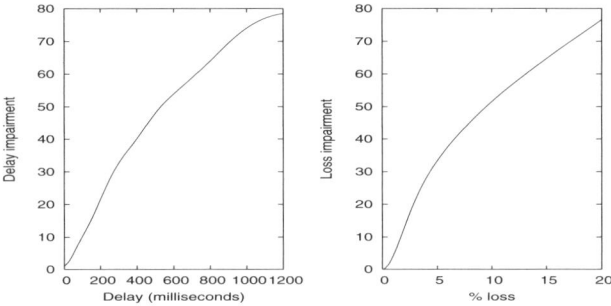

Fig. 2. Delay and loss impairment values

a speaker with a poor network connection who speaks for a very short time, say five seconds, in an otherwise good quality conversation, can cause the overall quality to be underestimated. We set t to be eight seconds in our implementation. This setting is long enough to reflect the time a person would take to form an opinion about the quality. Yet, it is short enough that averaging effects as noted above are minimized.

Long-term conference MOS is the mean of the short-term conference MOS values. Mean short-term MOS has been found to be a good predictor of long-term MOS [17].

4   Measurement Methodology

Our goal for the measurement study is to investigate if a dial-in conferencing solution is well suited for enterprise-grade audio conferencing on the Internet. In order to achieve this goal, we monitored and characterized the performance of the GoToMeeting conferencing solution using the Conference MOS metric. Our data set includes traces from end points around the world. Therefore, we believe that our results will likely be representative of other similar solutions.

4.1   VoIP Infrastructure

Our VoIP infrastructure consists of GoToMeeting [1], a collaboration application installed on End Points (EPs), and Voice Conferencing Bridges (VCBs) geographically distributed at various data centers. The GoToMeeting application allows users to collaborate online, give presentations and provide online trainings. A GoToMeeting user can participate in an audio conference either via VoIP, with a mic and headset, or dial in using a PSTN phone. PSTN gateways facilitate the mixing of PSTN and VoIP traffic.

The VCBs receive packetized audio from the EPs over RTP, perform audio mixing and transmit the mix to the EPs. The EPs initiate conferences on VCBs using SIP. VCB selection is performed using a proprietary algorithm. Closeness, in terms of network latency, to the organizer of a conference is one parameter in the algorithm. Note that a VCB located in a country will have connections from users outside that country if conference participants are situated abroad.

4.2 Data Sets

We collected five days of data from Aug 5-9, 2008 from a subset of VCBs that are part of our VoIP infrastructure. These VCBs are located in data centers on the east and west coasts of the United States. Our data set is made up of packet traces and VCB log files. The traces contain RTP headers and RTCP packets. We did not record the RTP payload in order to ensure user privacy. RTP headers have contributing source (CSRC) information, which gives the list of speakers encoded in the RTP packet. The speaker list is utilized to compute the Conference MOS. The RTCP packets carry bidirectional loss and delay information. The log files contain conference information, such as a conference's creation time and its lifetime.

Our data set contains 9394 VoIP flows from end points around the world. These flows connected to 7436 conferences. Given that a conference consists of multiple users, one would expect the number of VoIP flows to be substantially higher than conferences. In the case of these conferences, some users chose to call in via PSTN.

VoIP flows are encoded using *iSAC* [4], a wide-band codec (16Khz sampling rate) from Global IP Solutions. These flows totalled 231000 minutes of talk time. Figure 3 shows the arrival of VoIP connections and the creation of conferences per hour over the five day period. A conference is created on a VCB when a VoIP/PSTN flow arrives for that conference. PSTN arrivals are not shown on this graph. The arrivals follow a typical diurnal pattern with peaks during the day. The night-time load is about one third that of the day. As expected, less activity is seen over the weekends. The peak number of VoIP arrivals per hour in our data set is between 150 to 200 whereas the peak number of conferences created per hour is between 100 and 140. At certain times, such as seen in the friday section of the graph, the number of VoIP connections is only slightly higher than conferences, because some users in these conferences joined using their PSTN phones.

Figure 4 shows the lifetime of conferences and VoIP connections as a CDF. The median and average lifetimes of VoIP flows are 10 minutes and 24 minutes respectively, whereas the median and average lifetimes of conferences are approximately 30 minutes and 49 minutes respectively. The lifetime of a VoIP flow is typically shorter because a user might join or leave a conference at any time.

Birke et. al. [12] measured the average VoIP flow duration within an Italian ISP to be of much shorter duration (approximately 2 minutes). We attribute the difference to the application use-cases. The Italian ISP is mainly used for one-to-one social/business calls [12], whereas conferences in GoToMeeting involve users who collaborate online on a specific topic.

5 Traffic Analysis

Our goal is to study the MOS of conferences hosted on our VoIP infrastructure. We calculate a participant's conference MOS and *connection MOS* using the technique outlined in Section 3.2 and the E-model respectively. The Connection MOS is simply the MOS on the path between the participant and the Voice Conferencing Bridge (VCB). We calculate connection MOS because a participant's connection to the VCB is a key contributor to its conference MOS. Further, connection MOS is a good baseline for

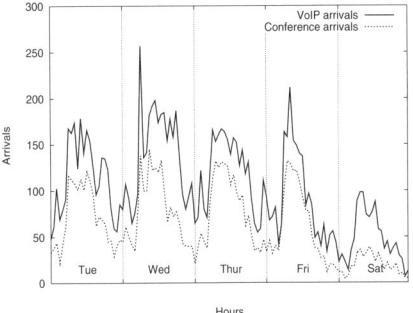

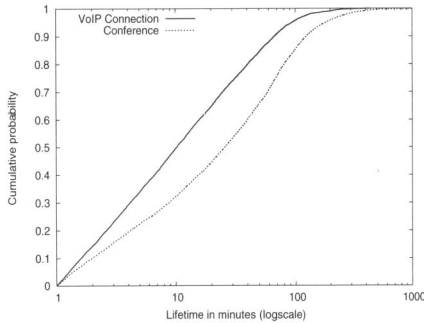

Fig. 3. Arrivals of VoIP connections and conferences

Fig. 4. Lifetime of VoIP connections and conferences

comparison purposes. Conference MOS and connection MOS higher than 3.5 is classified as good quality, whereas anything lower is undesirable.

We wrote a tool that computes these MOS values using: (1) conference details captured in log files; (2) loss and delay measurements in RTCP packets; and (3) speaker information from RTP packet headers.

5.1 Results

Figure 5 plots the average connection MOS value and average conference MOS value as a CDF. The average is computed over the lifetime of a participant's flow by taking the mean of the respective instantaneous MOS values, which are calculated every eight seconds (motivated in Section 3.2). Average MOS has been found to be a good predictor of long-term MOS [17]. This paper does not investigate variations in instantaneous MOS values. We plan to expore such variations as future work.

We make two observations from this figure. First, the median conference MOS is approximately 4.20, while the median connection MOS is about 4.30. Only 11.75% of participants have conference MOS less than 3.5. 11% of participants have connection MOS less than 3.5. *This result implies that a majority of participants, in general, experienced good quality for conferences hosted on the GoToMeeting VoIP infrastructure.*

Second, conference MOS appears lower than connection MOS. This is because the conference MOS depends not only on the participant's connection to the VCB, but also on the network connections of the speakers in the conference, one or many of which can adversely impact the participant's conference MOS.

In order to further explore the relationship between conference MOS and connection MOS, we plot in Figure 6 the conference and connection MOS values for hundred randomly selected participants. Points on the line $x = y$ indicate all connections with similar conference and connection MOS. Points above this line are cases with higher connection MOS than conference MOS. There are no points below this line because a participant's conference MOS can only be as good as its connection MOS. As can be seen in the graph, some participants have connection MOS higher than 3.5, but conference MOS less than 2.5. For such cases, the remaining participants in their respective

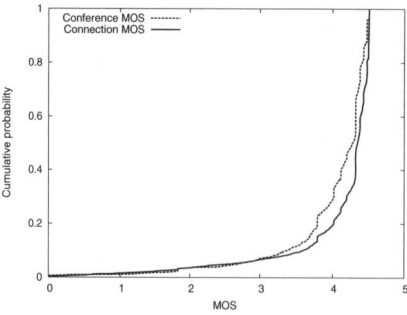
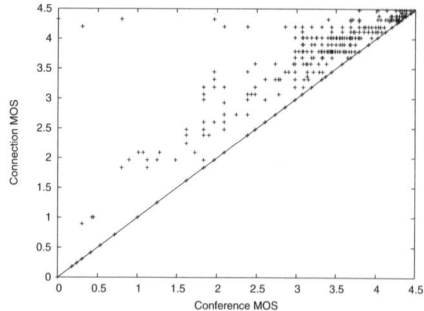

Fig. 5. Connection and conference MOS as a CDF

Fig. 6. Participant's connection and conference MOS as a scatter plot

conferences have likely very poor connections to the VCB and, as a consequence, impact the conference MOS in a severe manner.

These poor connections are a part of the 11% of participants with connection MOS less than 3.5 in Figure 5. Some of these connections could be from international participants. Figure 7 plots the CDF of the average connection MOS for all participants, but also segmented into participants within the US and ones outside the US. We observe that international participants have relatively poor quality connections compared to local participants. The median connection MOS for local participants is 4.39, while with international participants, it is slightly lower at 4.11. 16.74% of international participants had connection MOS less than 3.5 while only 9.19% of local participants had this low value. As such, the lower quality is to be expected given the transcontinental delay and loss on paths to the VCBs in the US. However, almost 10% of US participants also had connection MOS less than 3.5.

In order to understand why these connections had low MOS, in Figure 8, we plot the average delay and loss impairments as a scatter plot for all participants with connection MOS less than 3.5. The graph is divided into four quadrants. We observe that loss and delay contribute to MOS in varying proportions. For example, points in the second quadrant (Q2) indicate that high delay is the principal contributor to low MOS; points in quadrant four (Q4) show that high loss is the principal contributor, whereas points in the third quadrant (Q3) indicate that high loss and high delay contribute in approximately equal proportions; points in the first quadrant (Q1) indicate that although delay and loss were low individually, a combination of the two leads to bad audio quality. Our finding differs from that of other studies [12,14] which conclude that loss, not delay, is the main reason for poor quality. These studies were limited to backbone networks which are typically provisioned for high capacity; end-to-end paths are likely limited because of the "last-mile".

5.2 Discussion

Our results indicate that the Internet is suitable for enterprise-grade audio conferencing. While our results are specific to the VoIP infrastructure we monitored, we believe that our conclusion is applicable to other similar deployments.

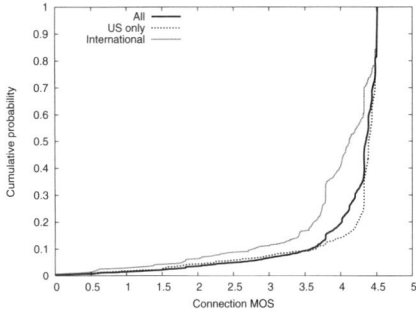

Fig. 7. Connection MOS for US and international participants

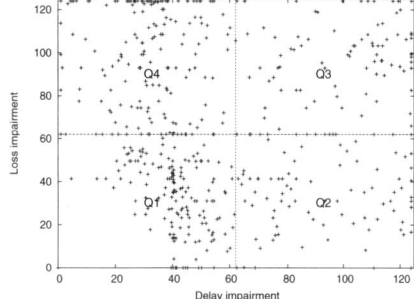

Fig. 8. Average impairment factors for participants with connection MOS less than 3.5

Participants with poor network connections will be the "weakest-links" in achieving good conference quality. Strategies to reduce their influence include the placement of VCBs in the proximity of such participants and to utilize sophisticated routing techniques [19]. For example, one strategy could be to deploy VCBs in an overlay [11] that covers hotspots around the world and to interconnect these VCBs with high bandwidth connections in order to mix participants that span hotspots. While such a strategy could alleviate some of the poor quality, it will likely fail to eliminate all such cases.

Therefore, we believe that in order to support enterprise-grade conferencing, a solution that allows users with poor network connections to dial in using their PSTN phones is a better alternative. The Internet backbone is provisioned for high capacity. It is conceivable that the "last-mile" is similarly provisioned in the near future. Until such time, enterprise-grade audio conferencing solutions should utilize a deployment strategy that combines VoIP and PSTN than a pure VoIP solution.

6 Conclusion

In this paper, we investigate if the Internet can support enterprise-grade audio conferencing. We use the proposed Conference MOS metric to study the performance of the GoToMeeting audio conferencing solution. Our results indicate that a majority of users utilizing this solution experienced good conference quality. A minority would have been better served if they had dialed in using PSTN phones. This leads us to conclude that a deployment strategy that supports VoIP and PSTN would serve users better than a VoIP only solution.

Conference MOS is useful for analyzing the quality of an audio conference. We believe it can help network operators monitor and troubleshoot their deployments. Furthermore, it can be used to proactively notify a user of his poor quality connection so that he can take corrective measures, such as calling in with a PSTN phone.

The results presented in this paper shed much needed light on the performance of audio conferencing in the Internet. As future work, we plan to expand our monitoring effort to include a larger set of conferencing bridges and VoIP flows. We also plan to study the transient behavior of VoIP flows and its impact on quality.

Acknowledgments

Albert Alexandrov, Bruce Brown, Robert Chalmers, Nitin Gupta, Alan Knight, Ashwin Sampath and Chris Scheiman provided valuable technical support for the paper. The Operations and IT teams helped with traffic collection and storage.

References

1. Citrix GoToMeeting, `http://www.gotomeeting.com`
2. Citrix GoToWebinar,
   `http://www.gotowebinar.com/webinar/pre/reliability.tmpl`
3. Frost Conferencing Research,
   `http://www.frost.com/prod/servlet/svcg.pag/ITCF`
4. Global IP Solutions iSAC,
   `http://www.gipscorp.com/files/english/datasheets/iSAC.pdf`
5. North American Audio Conferencing Services Markets,
   `http://www.mindbranch.com/North-American-Audio-R1-6595/`
6. Skype Conference Calling,
   `http://www.skype.com/business/features/conferencecall/`
7. Methods for Subjective Determination of Transmission Quality. In: ITU-T P.800 (1996)
8. E-model, a Computation Model for use in Transmission Planning. In: ITU-T G.107 (2002)
9. Mean Opinion Score (MOS) Terminology. In: ITU-T P.800.1 (2003)
10. Provisional Impairment Factor Framework for Wideband Speech Transmission. In: ITU-T G.107 Amendment 1 (2006)
11. Amir, Y., Danilov, C., Goose, S., Hedgvist, D., Terzis, A.: 1-800-OVERLAYS: Using Overlay Networks to Improve VoIP Quality. In: ACM NOSSDAV, Stevenson, WA (June 2005)
12. Birke, R., Mellia, M., Petracca, M.: Understanding VoIP from Backbone Measurements. In: IEEE Infocom, Anchorage, AL (May 2007)
13. Bolot, J., Crepin, H., Vega-Garcia, A.: Analysis of Audio Packet Loss in the Internet. In: ACM NOSSDAV, Durham, NH (April 1995)
14. Boutremans, C., Iannaccone, G., Diot, C.: Impact of Link Failures on VoIP Performance. In: ACM NOSSDAV, Miami, FL (May 2002)
15. Cole, R., Rosenbluth, J.: Voice over IP Performance Monitoring. In: ACM SIGCOMM Computer Communication Review, vol. 31, pp. 9–24 (April 2001)
16. Markopoulou, A., Tobagi, F., Karam, M.: Assessment of VoIP Quality over Internet Backbones. In: IEEE Infocom, New York, NY (June 2002)
17. Rix, A., Beerends, J., Kim, D., Kroon, P., Ghitza, O.: Objective Assessment of Speech and Audio Quality — Technology and Applications. In: IEEE Transactions on Audio, Speech, and Language Processing, vol. 14, pp. 1890–1901 (November 2006)
18. Rosenberg, J., Schulzrinne, H.: Models for Multi Party Conferencing in SIP. Internet Engineering Task Force (IETF), draft-ietf-sipping-conferencing-models.txt (January 2003)
19. Tao, S., Xu, K., Estepa, A., Fei, T., Gao, L., Guerin, R., Kurose, J., Towsley, D., Zhang, Z.: Improving VoIP Quality Through Path Switching. In: IEEE Infocom, Miami, FL (March 2005)

Peer-to-Peer

PBS: Periodic Behavioral Spectrum of P2P Applications

Tom Z.J. Fu[1], Yan Hu[1], Xingang Shi[1], Dah Ming Chiu[1], and John C.S. Lui[2]

[1] IE Dept., CUHK
{zjfu6,yhu4,sxg007,dmchiu}@ie.cuhk.edu.hk
[2] CSE Dept., CUHK
cslui@cse.cuhk.edu.hk

Abstract. Due to the significant increase of peer-to-peer (P2P) traffic in the past few years, more attentions are put on designing effective methodologies of monitoring and identifying P2P traffic. In this paper, we propose a novel approach to measure and discover the special characteristics of P2P applications, the periodic behaviors, from the packet traces. We call this the "periodic behavioral spectrum" (PBS) of P2P applications. This new finding, learning the characteristics of P2P traffic from a new angle, could enhance our understanding on P2P applications. To show the effectiveness of our approach, we not only provide justifications as to why P2P applications should have some inherent periodic behaviors, but also conduct hundreds of experiments of applying the approach on several popular P2P applications.

1 Introduction

There is a significant increase in P2P applications running over the Internet and enterprise IP networks during the past few years. These applications include P2P content distribution applications like BitTorrent, BitComet and eMule, and P2P streaming applications like Sopcast, PPLive, PPStream. Since P2P applications account for a large portion of total Internet traffic, it is important to correctly identify P2P traffic for traffic monitoring and network operations. However, existing approaches to classifying P2P traffic have well known drawbacks: Port-based method is ineffective since many P2P applications rarely use fixed port numbers. Payload signature-based method is more reliable, but constraints like hardware resource limitation, payload encryption and privacy and legal concerns make it ineffective. Hence, it is important to have a better understanding of the characteristics of P2P traffic and thus being able to effectively differentiate it from other conventional applications such as Web and FTP.

In this paper, we propose a novel approach, *Two Phase Transformation*, to measure and discover "periodic" behaviors of P2P applications. In addition, we provide justifications to why P2P applications should have some inherent periodic behaviors. To show the effectiveness of our approach, we carry out a number of experiments by applying this novel approach on several popular P2P applications (such as PPLive, PPStream, eMule etc.). Interestingly, the experimental

S.B. Moon et al. (Eds.): PAM 2009, LNCS 5448, pp. 155–164, 2009.
© Springer-Verlag Berlin Heidelberg 2009

results show that different frequency characteristics of P2P applications could form a periodic behavioral spectrum (PBS), which could be used as a new form of signatures to help to solve the monitoring and identifying problem. This is the main contribution of this paper.

In the rest of the paper, we first introduce three different periodic communication patterns and provide justifications why P2P applications should have these periodic behaviors (section 2). We propose our approach of how to discover the periodic behavioral patterns of P2P applications from packet traces (section 3). Then we show PBS we developed for a number of popular P2P applications, by individually doing experiments and applying our approach to observe their behaviors in isolation, and its application (section 4). Finally we discuss the related work (section 5) and conclusion (section 6).

2   Periodic Group Communication Patterns

Independent of the service type (e.g., file sharing, content streaming, or VoIP), a P2P application needs to form an overlay with other peers for reachability. In order to form and maintain this overlay, and often in the use of this overlay, peers inevitably exhibit periodic group communication.

We distinguish between two classes of periodic group communication patterns: (a) *control plane* - that used to form and maintain the overlay; (b) *data plane* - that used to multicast content. In P2P systems, especially those P2P systems performing application layer multicasting, there are basically two kinds of overlays formed:

- Structured overlays: This includes overlays with mesh-based topology, such as ESM[4], and tree-based topology such as NICE[3] and Yoid[6]. In ESM, group members periodically generate refresh messages and exchange their knowledge of group membership with their neighbors in the mesh. Similarly for tree-based topologies, peers also periodically refresh the overlay links so as to maintain the *soft state* information.
- Data-driven overlays: The classic example is BitTorrent[5]. In this case, the topology is more dynamic, driven by which neighbors have the right content needed by a peer. Such dynamic P2P systems are normally bootstrapped by a server known as the *tracker*. All peers may need to periodically update their information to the tracker for system health monitoring.

In both kinds of overlays, some active measurements may be used to optimize the efficiency of the overlay. For example, neighboring peers may periodically measure the distance (in terms of round trip time) between each other. In summary, these activities generate periodic group communication patterns.

2.1   Terminology for Behavioral Patterns

In this section, we describe three specific periodic group communication patterns that are common for many P2P applications. Note that these three patterns of periodic group communication behaviors are just examples to illustrate our methodology. The particular values of periodicity of different behaviors are application dependent. A given P2P application may exhibit one or more of such behaviors.

In doing so, we need to define some terminology in our framework. First, time is divided into discrete time intervals. The length of the time interval is quite critical in the ability to identify the periodic behavior, and needs to be carefully chosen. Unless we state otherwise, the length of the time interval is set as 1 second for all our experiments. For the host running the P2P applications (*target host*), it communicates with a number of neighbors. Such communications are organized into a sequence of flows, similar to the flows defined in Netflow, although the inactivity interval that starts and ends a flow is application behavior dependent. The start and end of a flow is indicated with a Start Event (SE) and an End Event (EE), each event has an associated time stamp.

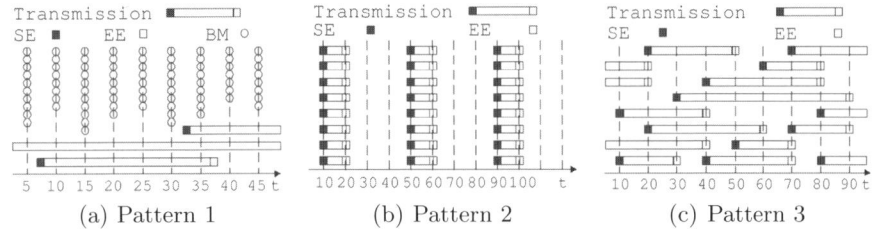

(a) Pattern 1   (b) Pattern 2   (c) Pattern 3

Fig. 1. Examples of three periodic communication patterns

Pattern 1 (Gossip of Buffer Maps): A popular type of P2P applications is P2P streaming based on a data-driven overlay. This includes live streaming applications such as Sopcast, PPLive, PPStream and P2P Video on Demand (VoD) streaming systems. Because of the data-driven approach for forming and maintaining the overlay, all these applications rely on gossip of buffer maps to maintain active links between peers. Typically, one or two packets are enough to present a buffer map.

For P2P Live Streaming systems, because peers only store the streaming content in their RAM (much smaller size compared with VoD) and remove the content as soon as it is played back, the buffer map information changes rather quickly. Each peer needs to periodically exchange its buffer map with neighbors to optimize the scheduling of content exchange to ensure good playback performance. The buffer map exchange period is as short as 5 seconds for some cases. Note, each peer must exchange this information with all its neighbors although it only exchanges content with a subset of these neighbors. For P2P VoD systems, peers also exchange buffer maps with their neighbors periodically, although the period may be longer. Figure 1(a) is an illustration of the traffic pattern for P2P Live Streaming systems. In this figure, we have three flows, e.g., there is a flow which starts right after $t = 5$ and ends between $[35, 40]$. The figure also shows the periodicity property. For example, every 5 seconds, this node sends out some packets to its neighbors (say at $t = 5$, this node sends out 10 packets describing its buffer maps to its neighbors).

Pattern 2 (Content flow control): The second pattern occurs in the data plane. For streamed video content, it often happens that peers download at a higher speed than the playback rate, behaving like file downloading. Although this is good for these peers, the content provider actually prefers the peers download at the pace of playback, to ensure all the peers stay around to help the server in distributing content, rather than watch the content off-line. Content providers thus implement various mechanisms to make peers continue to contribute. One way is to make peers periodically send keep-alive messages to a tracker when they are watching the video, even after the whole video has finished downloading (e.g. in the VoD case[7]). Another way is to perform the Pre-Downloading Control (PDC), which is a form of content flow control to make the download rate match with the playback rate. Such flow control often results in alternating bursts of download activities and sleep periods, as illustrated in Figure 1(b).

Pattern 3 (Synchronized Link Activation and Deactivation): It is well-known that BitTorrent implements the *tit-for-tat* mechanism to provide incentives for peers to serve each other. The third pattern of periodic group communication behavior is a direct consequence of how BitTorrent-like protocols might implement the tit-for-tat mechanism. As described in [5,10], each peer uses two timers (10 seconds and 30 seconds) to decide whether to choke and optimistically unchoke neighboring peers, respectively. This results in the synchronization of Start Events (SE) and End Events (EE) at the beginning of the time intervals, as illustrated in Figure 1(c).

3   Discovering Periodic Behavioral Patterns

In this section, we describe the approach of how to discover the periodic behaviors of P2P applications, especially the periodic patterns discussed in Section 2.

The overview of the approach is as follows. First, we run a particular P2P application and collect the application's packet trace in a controlled environment where all other network applications are disabled from the target host. While doing so, we only collect packet header information.

Second, we feed the packet trace into three independent and parallel analyzing processes. For each analyzing process, there are two transformation phases. The first one is the transformation from packet-trace to discrete-time sequence, or sequence generator (SG). Three different sequence generators are specifically designed to extract those three periodic patterns. The second transformation phase is the same for all three analyzing processes. It transforms the time-domain sequence to frequency-domain sequence. In this phase, we first apply the Auto-correlation Function (ACF) then the Discrete Fourier Transform (DFT[1]).

---

[1] In our implementation, Fast Fourier Transform (FFT) is applied.

Finally, we analyze the frequency-domain results derived by ACF and DFT to identify periodic characteristics. In the following section, we present the three sequence generators in detail along with some empirical results.

3.1 Sequence Generators

SG1: Time Series for the Gossip Pattern. Recall that in our basic model, time is divided into intervals, of length T. $X_{in}[i]$ and $X_{out}[i]$ denote time series generated by SG1, where $X_{in}[i]$ represents for the number of source hosts sending data to the target host during the ith interval; and $X_{out}[i]$ is correspondingly the number of destinations which are receiving data from the target host. When the target host is engaged in gossiping, $X_{in}[i]$ and $X_{out}[i]$ represent the number of neighbors gossiping with the target host over the time interval i.

Then the ACF is applied on $X_{in}[i]$ and $X_{out}[i]$ respectively. We denote $r_{Xin}(n)$ and $r_{Xout}(n)$ the result sequences. Finally, we apply DFT on $r_{Xin}(n)$ and $r_{Xout}(n)$ and derive the frequency-domain results denoted by $\mathbf{R_{Xin}}(\frac{k}{N})$ and $\mathbf{R_{Xout}}(\frac{k}{N})$. Since the ACF and DFT are basic functions in signal processing, the definition and detailed explanation of them can be found in many books and articles, such as[13]. Here we just give the basic formulas. The ACF and DFT are described in Eq. (1) and (2) where $X(i)$ is any input time-domain sequence (e.g., $X_{in}[i]$ or $X_{out}[i]$) and N is the sequence length of $X(i)$.

$$r(n) = \frac{1}{N-n} \sum_{i=1}^{N-n} X(i)X(i+n). \tag{1}$$

$$\mathbf{R}(k) = \sum_{n=0}^{N-1} r(n)e^{-\frac{2\pi i}{N}kn} \quad \text{where } k \in [0, N-1]. \tag{2}$$

The sequence, $\mathbf{R}(0), \mathbf{R}(1), \ldots, \mathbf{R}(N-1)$ is a sequence of N complex numbers (see [13]). For discovering the periodic behavioral patterns, it is sufficient for us to take the modulo of $\mathbf{R}(k)$ to get the magnitude of each frequency component.

For example, Figure 2 shows $X_{out}[i]$ and its ACF and FFT transformations for a PPLive streaming session ($N = 200$). In Figure 2(c), we observe that there

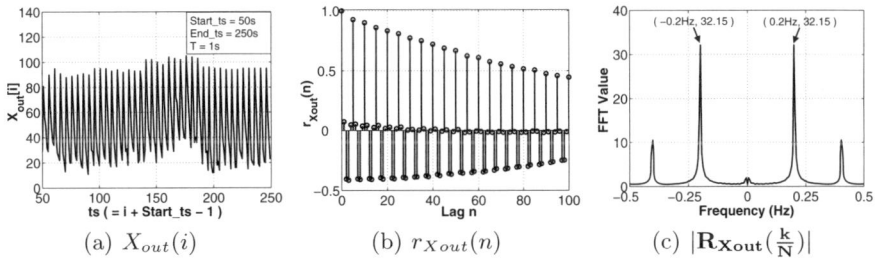

(a) $X_{out}(i)$ (b) $r_{Xout}(n)$ (c) $|\mathbf{R_{Xout}}(\frac{k}{N})|$

Fig. 2. $X_{out}[i]$, ACF and FFT transformation results for a PPLive streaming session

is a frequency pulse at $f = 0.2$Hz, which means sequence $X_{out}[i]$ has a 5-second periodic characteristic and reveals the periodic gossip pattern.

SG2: Time Series for Content Flow Control Pattern. Recall that the content flow control traffic pattern (Figure 1(b)) is about the rate a target host is sending or receiving content from all its neighbors. We represent these as two time-domain sequences, $Y_{in}[i]$ and $Y_{out}[i]$.

The procedure of SG2 is similar to SG1. $Y_{in}[i]$ and $Y_{out}[i]$ are used to accumulate in and out data transmission rate during the ith interval separately, rather than flow count in and out of the target host. After the time-domain sequences are generated, ACF and FFT are applied.

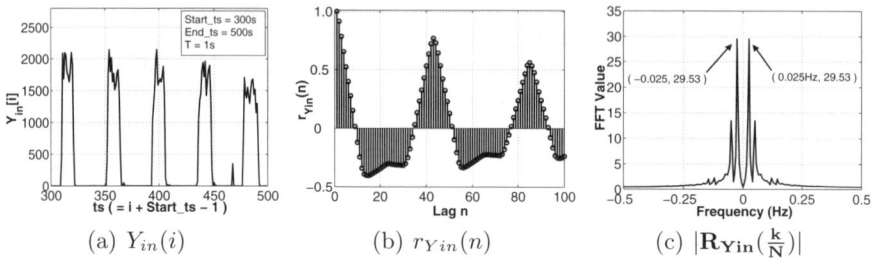

(a) $Y_{in}(i)$ (b) $r_{Y in}(n)$ (c) $|\mathbf{R_{Y in}}(\frac{k}{N})|$

Fig. 3. $Y_{in}[i]$, ACF and FFT transformation results for a PPStream VoD session

We illustrate an example of PPStream VoD session ($N = 200$). Figure 3 shows $Y_{in}[i]$ and its ACF and FFT transformation results. From the frequency pulse at $f = 0,025$Hz as shown in Figure 3(c), it becomes apparent that PPStream VoD session executes the PDC mechanism in every 40 seconds.

SG3: Time Series for Synchronized Start and Finish of Flows. In BitTorrent-like applications, due to the periodic choking and optimistic unchoking mechanism, the occurrences of the data transmission Start Event (SE) and End Event (EE) will also have the periodicity (Refer to Figure 1(c)). This time, the results will be accumulated in time-domain sequences $Z_{in}[i]$ and $Z_{out}[i]$. The algorithm of SG3 is slightly more complicated than the first two algorithms.

There are three steps of this algorithm. In the first step, all the packets in the input packet trace are reorganized into flows according to their five-tuple information {srcIP, srcPort, dstIP, dstPort and protocol} and then sorted in the ascending order by their *Time Stamp* (TS). Flows destined for the target host are *in-flows*; others are *out-flows*.

In the second step, all the flows are divided into subflows, each subflow with its distinctive SE and EE. Each subflow should correspond to content exchange

TS:	1	2	3	4	5	6	7	8	9	10	11	12
Flow:	P1			P2		P3					P4	P5
Subflow:				Subflow1							Subflow2	

Fig. 4. An example of how to separate a flow into subflows (interval_threshold $= 3$)

between the target host and one of its neighbors. The rule for marking the beginning and end of subflows is that the time interval of any two consecutive packets of the same flow should not be larger than the given parameter interval_threshold. This is like the *inactivity timer* in Netflow[1]. Figure 4 gives a simple example. In the end, all the triggered events are sorted into ascending order of TS.

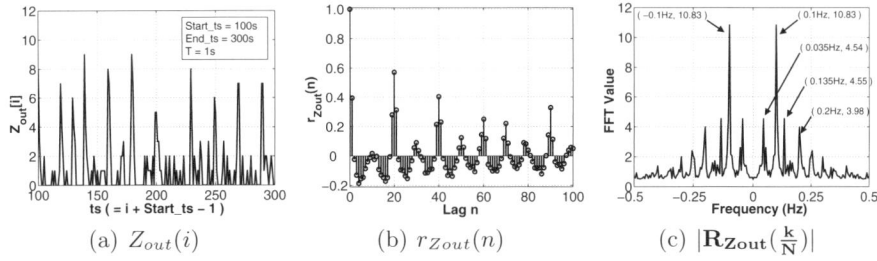

(a) $Z_{out}(i)$ (b) $r_{Zout}(n)$ (c) $|\mathbf{R_{Zout}}(\frac{k}{N})|$

Fig. 5. $Z_{out}[i]$, ACF and FFT transformation results for a BitTorrent session

The third step is similar to the algorithms in SG1 and SG2. The time-domain sequences $Z_{in}[i]$ and $Z_{out}[i]$ represent the total number of SEs and EEs of all the *in-flows* and *out-flows* triggered in the ith time interval respectively.

Figure 5 shows $Z_{out}[i]$, ACF and FFT transformation results for a BitTorrent session ($N = 200$, packet_number_threshold = 10, interval_threshold = 4). Theoretically, the value of interval_threshold, which determines the start and end of subflows, is likely to affect the frequency characteristics of the sequence. The larger the value is, the fewer number of events will be triggered and there is a higher probability that the application frequency will be buried by noise. So, interval_threshold should take a relatively small value.

In 5(c), we observe that there are four frequencies with large FFT values. In fact, the frequency points $f_1 - 0.035$Hz and $f_2 - 0.1$Hz are the frequency characteristics caused by choking (every 10 second) and optimistic unchoking (every 30 second). The remaining two frequencies, ($f_3 = 0.135$Hz and $f_4 = 0.2$Hz) are the harmonic frequencies, which are the linear combination of the basic frequencies 0.1Hz and 0.035Hz.

4   Frequency Characteristics of Popular P2P Applications

In this section, we present the experimental results of the frequency characteristics of several popular P2P applications. These characteristics are derived from the frequency-domain analysis (as discussed in Section 3) of the real packet traces that we captured in a controlled environment.

Packets are captured using Wireshark[2] and each experiment lasts for 30 minutes. When we run each P2P application, we *turn off* all other network applications running on the target machine. In Table 1, we list the frequency characteristics of 15 popular P2P applications and periodic behavioral spectrum (PBS). We also selectively plot the FFTs of these applications in Figure 6. For

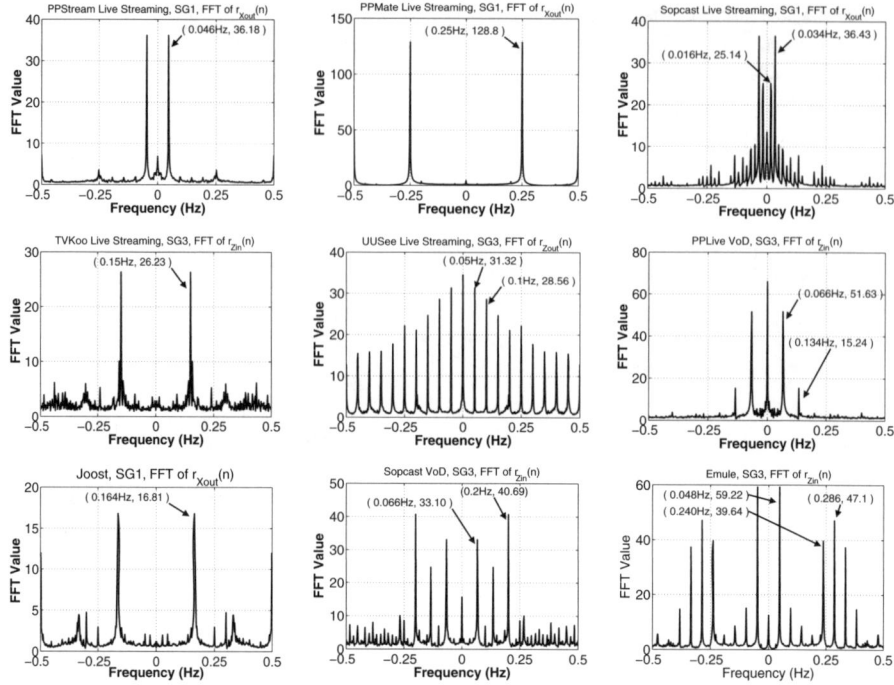

Fig. 6. FFT results of nine selective P2P applications

each P2P application, we ran multiple experiments (with different settings) and analyzed the resulting packet trace to double check whether its frequency characteristics showed up every time. For example, for PPLive streaming application, we repeated experiments over six times on different channels of three popularity levels, at least twice per level: the most popular, moderately popular and the least popular. The results confirmed that the traffic exhibited the same frequency characteristics irrespective of a popularity level. In addition, for some of the applications (e.g., PPlive, PPStream), we also carry out the measurement under another controlled environment in which the computer accessed to the Internet was through ADSL instead of the campus Ethernet. Results showed that we could still find the frequency characteristics but the magnitude of FFT values at those frequency points were a bit smaller.

Table 1 shows that most P2P applications have unique fundamental frequencies. Exceptions are PPLive Streaming and TVAnt Streaming, which interestingly share the same frequency characteristics. We believe that the PBS shown in Table 1 can serve as a new form of signatures for classifying P2P applications from mixed packet trace.

Let us present some identification results from the mixed traffic traces collected from our department gateway. The packet header information of each packet was required by the PBS-based approach, but the payload was only used for validation. The measurement duration was two days. After applying the

Table 1. The PBS of 15 popular P2P applications

P2P Application Name	TCP/ UDP)	In or Out	Fundamental Frequency(Hz)	Harmonic Frequency(Hz)	Effective SGs
PPStream Streaming	TCP	Both	0.046		SG1, SG2, SG3
PPMate		Both	0.25		SG1
Streaming		Out	0.25		SG2
PPLive Streaming	Both	Both	0.2	0.4	SG1, SG2
TVAnt Streaming	Both	Both	0.2	0.4	SG1, SG2
Sopcast		Out	0.016, 0.034	0.066	SG1
Streaming	UDP	Both	0.016, 0.034	0.066, 0.1	SG3
TVKoo Streaming	UDP	Both	0.15		SG1, SG3
UUSee	UDP	Both	0.05	0.1, 0.15, 0.2, 0.25	SG3
Streaming	TCP	Both	0.1	0.2, 0.3, 0.4	SG3
TVU Streaming	UDP	In	0.034, 0.066	0.1	SG1, SG3
PPStream VoD	UDP	Both	0.024	0.048, 0.072	SG1, SG2, SG3
		Out	0.1	0.2	SG1
PPLive VoD	UDP	Both	0.066	0.134	SG3
Joost	UDP	Both	0.164	0.328	SG1, SG2
UUSee VoD	UDP	Both	0.05	0.1, 0.15	SG1, SG2, SG3
Sopcast VoD	UDP	Both	0.066	0.134, 0.2	SG3
eMule	UDP	Both	0.048	0.192, 0.24	SG3
BitTorrent	TCP	Both	0.034, 0.1	0.134	SG1, SG3

PBS-based identification approach, four P2P applications were found and they were PPStream, PPLive, eMule and BitTorrent. We then used a combined method including payload signature checking and manual analysis for validation. The validation results showed that the heuristic approach worked well (with 100% accuracy). Although the PBS-based approach is a prototype in the current stage, we believe that the application of the PBS is promising and valuable.

5   Related Works

The tremendous growth of P2P traffic has drawn much attention from researchers. Several studies emphasize on identification of P2P traffic, such as the signature-based payload method in [14] and identifying by transport layer characteristics [8]. Recently, a novel approach named BLINC is proposed by Karagiannis et al. [9]. Although both BLINC and our approach are host-level methods, there is a significant difference between them. BLINC focuses on the behaviors of a host's connection patterns (spatial behaviors) while ours focuses on the periodic behaviors of a given host (temporal behaviors). Moore et al. in [11] apply Bayesian analysis techniques to categorize traffic by application. They also apply FFT to build discriminators [12], but the difference is that their method focuses on each single flow, i.e., applying FFT on the interarrival time of packets belonging to a single flow. Our approach, on the other hand, focuses on the host-level behaviors and we inspect the periodicity of all flows related to the same host.

6   Conclusion

In this paper, we first introduce three periodic communication patterns that most P2P applications have and provide concrete justifications. Then we present a novel approach called *Two Phase Transformation* to measure and discover these periodic behaviors of P2P applications. We carry out a large number of experiments applying this approach on several popular P2P applications (such as PPLive, PPStream, eMule etc.), and show the results of different frequency characteristics of P2P applications. These frequency characteristics can form a periodic behavioral spectrum (PBS), which can be used as a new form of signatures to help to monitor and identify P2P traffic.

Acknowledgments. We thank the reviewers and our shepherd for providing very helpful technical comments and editing help. This work is partially supported by NSFC-RGC grant N_CUHK414/06 from the Hong Kong government.

References

1. NetFlow, http://www.cisco.com/web/go/netflow
2. Wireshark, http://www.wireshark.org/
3. Banerjee, S., Bhattacharjee, B., Kommareddy, C.: Scalable application layer multicast. In: Proc. ACM SIGCOMM 2002 (August 2002)
4. Chu, Y., Rao, S.G., Zhang, H.: A case for end system multicast. In: Proc. ACM Sigmetrics 2000 (2000)
5. Cohen, B.: Incentives build robustness in bittorrent (May 2003), http://bitconjurer.org/BitTorrent/bittorrentecon.pdf
6. Francis, P.: Yoid: Extending the multicast internet architecture. White paper (1999)
7. Huang, Y., Fu, T.Z.J., Chiu, D.M., Lui, J.C.S., Huang, C.: Challenges, design and analysis of a large-scale p2p-vod system. In: Proc. ACM SIGCOMM 2008 (2008)
8. Karagiannis, T., Broido, A., Faloutsos, M., Claffy, K.: Transport layer identification of p2p traffic. In: Proc. IMC 2004 (2004)
9. Karagiannis, T., Papagiannaki, K., Faloutsos, M.: Blinc: Multilevel traffic classification in the dark. In: Proc. ACM SIGCOMM 2005 (2005)
10. Legout, A., Liogkas, N., Kohler, E.: Clustering and sharing incentives in bittorrent systems. In: Proc. ACM Sigmetrics 2007 (June 2007)
11. Moore, A.W., Zuev, D.: Internet traffic classification using bayesian analysis techniques. In: Proc. ACM Sigmetrics 2005 (2005)
12. Moore, A.W., Zuev, D., Crogan, M.: Discriminators for use in flow-based classification. Technical report, Intel Research, Cambridge (2005)
13. Oppenheim, A.V., Schafer, R.W., Buck, J.R.: Discrete-time signal processing, 2nd edn. Prentice-Hall, Englewood Cliffs (1999)
14. Sen, S., Spatscheck, O., Wang, D.: Accurate, scalable in-network identification of p2p traffic. In: Proc. WWW 2004 (2004)

Measuring Mobile Peer-to-Peer Usage: Case Finland 2007

Mikko V.J. Heikkinen, Antero Kivi, and Hannu Verkasalo

TKK Helsinki University of Technology, P.O. Box 3000, FI-02015 TKK, Finland
{mikko.heikkinen,antero.kivi,hannu.verkasalo}@tkk.fi

Abstract. We study the development of mobile peer-to-peer (MP2P) traffic and the usage of MP2P applications in Finland during 2005-2007. Research data consists of 1) traffic traces measured from three Finnish GSM/UMTS networks covering the Internet-bound mobile data traffic generated by 80-90% of Finnish mobile subscribers (N > 4,000,000), and 2) usage log files collected with a dedicated Symbian handset monitoring application (N = 579). In the traffic trace measurement, we notice almost zero P2P file sharing traffic for handsets, but 9-18% of unidentified traffic, part of which possibly being P2P traffic. Potentially a notable growing trend for computer-based P2P file sharing traffic is visible in GSM/UMTS networks, BitTorrent and eDonkey being the most popular protocols. In the panel study, only Fring, a client to several P2P-based communication services, has significant usage and data volume levels.

Keywords: Measurement, mobile, peer-to-peer, traffic, usage.

1 Introduction

Peer-to-peer (P2P) traffic has been rapidly growing in the Internet during past years [1]. On the other hand, mobile devices and laptops are increasingly using GSM/UMTS (Global System for Mobile communications / Universal Mobile Telecommunications System) connections to access the Internet [2]. Our aim is to study the development of mobile P2P (MP2P) traffic and the use of MP2P applications in Finland during 2005-2007 from two perspectives: 1) by analyzing traffic traces measured at the GSM/UMTS networks of three major Finnish mobile operators, and 2) by investigating results from a panel study conducted with a handset monitoring application running in Nokia Symbian S60 operating system. In the traffic trace measurement, we identify P2P file sharing traffic based on TCP/UDP port numbers and use TCP fingerprinting to differentiate between computers and handsets. In the handset monitoring, we identify MP2P applications based on logged application name, and record the number of usage sessions and transferred data volume. Our approach is restricted to the end-user viewpoint.

No commonly accepted definition for MP2P exists. According to Steinmetz and Wehrle [3], P2P is "a system with completely decentralized self-organization and resource usage". Androutsellis-Theotokis and Spinellis [4] attribute the following

S.B. Moon et al. (Eds.): PAM 2009, LNCS 5448, pp. 165–174, 2009.

characteristics to P2P systems: distribution, node interconnection, self-organization, resource sharing, adaptation to failures and transient populations, maintenance of acceptable connectivity and performance, and absence of global centralized servers or authority. We use the term "mobile" to describe both laptop computers and handsets with a data transfer connection to a GSM/UMTS mobile network. We cover both P2P systems consisting only or partially of mobile nodes, and mobile clients to P2P systems consisting only of fixed nodes. Of the handset applications we analyzed, only SymTorrent acts as a fully functional peer in a P2P system, other applications act as clients to P2P systems, i.e. relay traffic via intermediating peers or servers without implementing full peer capabilities.

Previous traffic measurement studies on MP2P in GSM/UMTS mobile networks are concentrated on the evaluation of using specific applications in limited test scenarios [5]-[9]. According to the best of our knowledge, our study is the first holistic examination of MP2P usage in GSM/UMTS mobile networks.

Our paper is structured as follows: first, we conduct a brief literature study on previous research. Then, we present our research methods, and the results of our measurements. Finally, we conclude our findings.

2 Previous Research

P2P traffic can be identified using various alternative methods. The simplest method is to investigate the port numbers in TCP or UDP packet headers [10], [11]. Some P2P applications use static port numbers to relay traffic, making their usage analysis fairly straightforward. However, an increasing number of P2P applications randomize or let their users randomize the port numbers in use. Therefore, in some studies various statistical methods have been developed to identify P2P traffic [12]-[16]. Other studies identify application-specific signatures in packet payload [17]-[19]. Guo et al. [20] analyze the log files generated by centralized components of a P2P network.

The point of measurement varies significantly in previous P2P measurement studies. Some analyzed a border point of an Internet Service Provider (ISP) network [10], [11], [13]. Others investigated a border point of an academic network [15], [17], [18]. Henderson et al. [19] collected packet-level traces and syslog messages from several access points to an academic Wireless Local Area Network (WLAN). Wu et al. [16] obtained traces from a company providing a streaming service based on a P2P network.

The reported metrics in the studies can be grouped into several categories. Traffic-based metrics include bandwidth consumption and traffic volume [11], [18], [19]; connection and session durations and latencies [10], [11], [18]; packet and flow level distributions [15], [16]; traffic patterns over time [10], [11], [13], [16]-[20]; and upstream and downstream traffic comparisons [10], [11] [16], [19]. Peer-related metrics consist of geographical distribution of peers [10]; number of downloads, uploads and shares by peers [8]; number of peers over time [14], [16], [20]; and peer connectivity and locality [10], [11].

3   MP2P Usage Measurements

3.1   TCP/IP Traffic Measurements

We use TCP/IP traffic measurements to collect traces of IP traffic in GSM/UMTS mobile networks. The measurements were conducted simultaneously at the networks of three major Finnish GSM/UMTS mobile network operators during two weeks in September-October 2005, 2006, and 2007. The measurements took place at a point between the Gateway GPRS Support Node (GGSN) and the Internet, at each of the three networks measured. In total, the measurements included the Internet-bound packet-switched data traffic of approximately 80-90% of Finnish mobile subscribers, i.e. over 4,000,000 subscribers. The measurements resulted in traces with the packet headers of Internet and transport layer protocols, whereas all application layer proto-col headers were sanitized from the data.

The underlying operating systems of the end-user devices generating the traffic, for example Symbian and Windows, are identified by using a method called TCP finger-printing [21]. Different operating systems are recognized by identifying idiosyncra-sies in the implementation of their respective TCP/IP stacks. No application layer data is needed, as the method only uses certain TCP and IP header fields. Operating sys-tem identification using TCP fingerprinting is a fairly reliable method, but it leaves some of the traffic unidentified.

Different application protocols are identified from the traffic traces using transport protocol (TCP, UDP) port numbers. Furthermore, the analysis of P2P traffic by Karagiannis et al. [22] is used to classify specific port numbers as P2P traffic. Port-based identification of P2P traffic has some limitations. First, as many P2P applica-tions select the used port numbers dynamically, identification based on the default port might leave a lot of traffic unidentified. Second, some P2P applications also use port 80 (HTTP) to masquerade their traffic as web traffic in order to, for example, pass simple firewalls or gain higher priority. Due to the limitations of the process, only the use of the following file sharing protocols is identified: BitTorrent, Direct Connect, eDonkey, FastTrack, Gnutella and Napster. Several applications can use the protocols and their variations.

Table 1. TCP and UDP port numbers used in P2P file sharing protocol identification

	TCP	UDP
BitTorrent	6881-6889	-
Direct Connect	411, 412	411, 412
eDonkey	4661, 4662	4665
FastTrack	1214	1214
Gnutella	6346, 6347	6346, 6347
Napster	6699-6702	6257

Table 1 summarizes the TCP and UDP port numbers used for identification. We also depict traffic generated by web and email protocols with the following TCP port numbers: HTTP (80), HTTPS (443) and HTTP alternate (8080) for web, and SMTP (25), POP3 (110), IMAP (143), IMAP/SSL (993) and POP3/SSL (995) for email.

3.2 Handset Monitoring

We utilize a handset-based research method in collecting data from end-users [23]. The method provides data on the actual usage of mobile applications and services, as the measurements are conducted directly in the device. End-users participating in the study install a client application on their device having a Nokia Symbian S60 operating system. The client runs as a background process, observing user actions and storing usage data into memory. The data collected includes application usage, data sessions, communication activities, memory status and Uniform Resource Locator (URL) traces, among others. The data is collected at the level of events and sessions, including accurate time stamps and identifiers for participating end-users. The data is transmitted daily to centralized servers for analysis. The method is deployed in controlled panel studies, to which typically a few hundred panelists are recruited.

The panel lasted for 1-2 months between November 2007 and January 2008. The panelists (i.e. users participating in the study) are provided with €20 vouchers as a compensation to potential data transfer costs they have to bear due to research and are entered to prize draws.

The main shortcoming of the method is the adverse selection of panelists. Typically technologically enthusiastic persons or people motivated by the prize draws participate in this kind of research panels. In addition, the Symbian device penetration is still well below 20% in the Finnish market [2], therefore the panelists could potentially be characterized as early-adopter users.

The panelists are recruited from the subscriber bases of three major Finnish mobile operators, targeting only consumer customers. 579 panelists from whom at least three weeks of usage-data is collected are included in the dataset. 44% of panelists are using Nokia S60 3rd edition devices, and 56% use older 2nd edition devices. 25% of the panelists have WLAN functionality in their handsets. 79 % of the panelists are male. The most dominant age groups are 20-29 years (38%) and 30-39 years (30%). Most panelists (77%) are employed. Over half (58%) of the panelists have a usage based data plan, the rest have a quota based plan (31%) or a flat rate plan (11%).

In the panel measurements, we identify four MP2P applications: Fring, iSkoot, MobileMule and SymTorrent. We have no knowledge whether the applications were installed by the users before or during the panel. Also, we do not know which applications the user had installed but did not use during the panel. We identify the applications by analyzing a list containing all the applications the panelists had used. The decision whether to include an application for detailed analysis is based on its recorded name. The detailed analysis consists of an Internet search, the aim of which is to determine whether the application can be considered a MP2P application. If we had used a different classification scheme, for instance included push-to-talk and instant messaging applications which use servers for both control and media traffic, we would have classified significantly more applications as MP2P applications.

4 Results

4.1 TCP/IP Traffic Measurement Results

According to the data provided by the mobile operators to Statistics Finland [24], [25], the volume of packet-switched data traffic in Finnish mobile networks has been growing rapidly during the recent years (2005: 34,000 GB, 2006: 100,000 GB, 2007: 500,000 GB). The rapid growth in data traffic volumes is partly explained by increased penetration of UMTS capable handsets [2], expansion of UMTS network coverage to smaller cities, and High-Speed Downlink Packet Access (HSDPA) upgrades to mobile networks. These developments have been accompanied by introduction of alternative flat rate mobile data subscriptions and heavy marketing of data cards and Universal Serial Bus (USB) data modems.

Our results from the measurement weeks in Falls 2005, 2006, and 2007 are consistent with the yearly figures reported by Statistics Finland. A fourfold increase in overall traffic volumes was observed between Falls 2005 and 2006, as both computer and handset (Symbian) originated traffic grew in proportion. However, between Falls 2006 and 2007 the traffic by computers grew by a factor of fourteen, whereas the growth of handset traffic was more moderate and merely tripled. The obvious difference between our results and the figures from Statistics Finland is explained by the fact that traffic volumes in Finnish mobile networks started to grow rapidly in late Summer 2007, likely due to the operators' aggressive marketing of flat rate mobile broadband subscriptions bundled with HSDPA-capable USB dongles for laptop computers. Overall, our measurements show that the relative share of computer traffic in mobile networks has grown from 70-75% in 2005-2006 to over 90% in 2007, whereas the traffic share of handsets running Symbian operating system has dropped from around 15% to about 4% of all traffic.

The profile of handset and computer traffic by application protocol is presented in Fig. 1 and 2. Clear differences in the application protocol profile of computers and handsets can be observed. Handset traffic is dominantly web browsing, whereas the share of email is significant albeit decreasing in relative terms. Other protocols, especially P2P protocols, are marginal traffic-wise. For computers, the relative share of web browsing and email traffic have been decreasing at the expense of the traffic that could not be identified, which amounts to almost 60% of all traffic in the latest traces. A small share (4-5%) of P2P traffic is also observed.

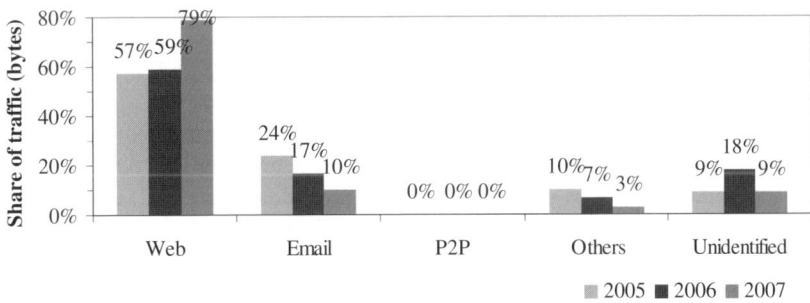

Fig. 1. Handset traffic by application protocol

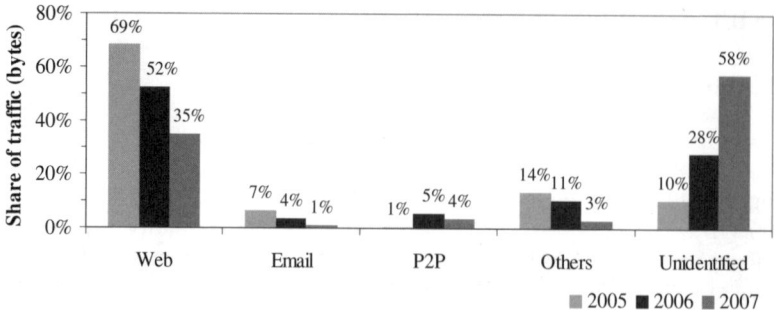

Fig. 2. Computer traffic by application protocol

The notable growth of unidentified traffic merits a further comparison with other traffic classes. Moreover, as P2P protocols are among the protocols typically using non-default port numbers, a first hypothesis would be that some of the unidentified traffic would in fact be P2P traffic. The diurnal distribution of computer traffic by application protocol for the measurement done in 2007 is presented in Fig. 3.

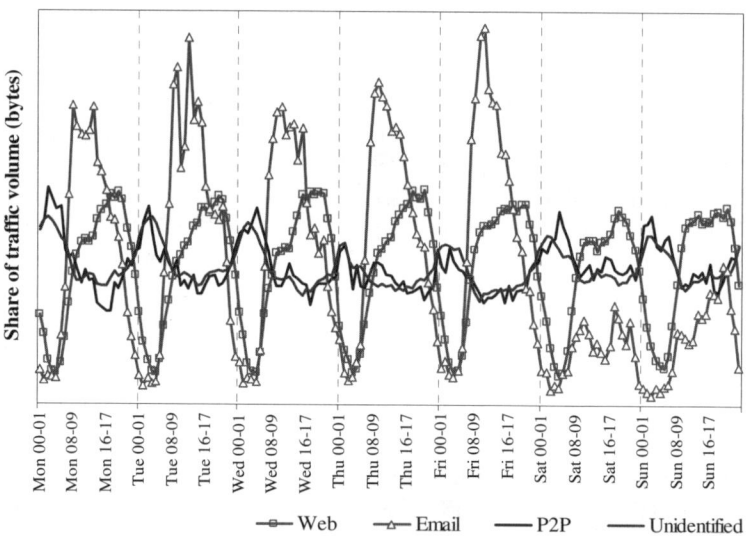

Fig. 3. Diurnal distribution of computer traffic by application protocol in 2007

The distributions of web and email traffic clearly reflect patterns of human behavior, i.e. high activity throughout waking hours and dramatically decreasing usage during the night. Moreover, the use of some applications (e.g. email) concentrates on office hours during the working days, whereas others (e.g. web browsing) are more free-time oriented. On the contrary, both P2P and unidentified traffic follow very similar patterns through the week, and the peak hours for both traffic types occur during the night.

Typically in fixed access networks P2P traffic is uniformly distributed over the whole day [1], as large media files get downloaded at the limit of available capacity without the need for any human input. In mobile networks, phone calls and non-P2P data usage consume most of the network capacity during the day, but the remaining capacity is consumed by P2P applications during the night. Further investigation of the traffic traces shows that the share of uplink traffic for P2P traffic (58%) is similar to unidentified traffic (52%), and significantly higher than for other applications (e.g. email: 32%, web: 13%). All this suggests that the true share of P2P traffic in the mobile network is considerably higher than the level proposed by pure port based identification.

A breakdown of P2P traffic by protocol is presented in Fig. 4. BitTorrent displays a growing trend; Direct Connect, Gnutella and Napster exhibit a decreasing trend; and eDonkey shows variation. However, as unidentified traffic potentially consists mostly of P2P traffic, the protocol profile is a result of the identification method, not of the true usage. In other words, the portion of traffic using non-standard port numbers is unidentified, therefore potentially biasing the distributions significantly.

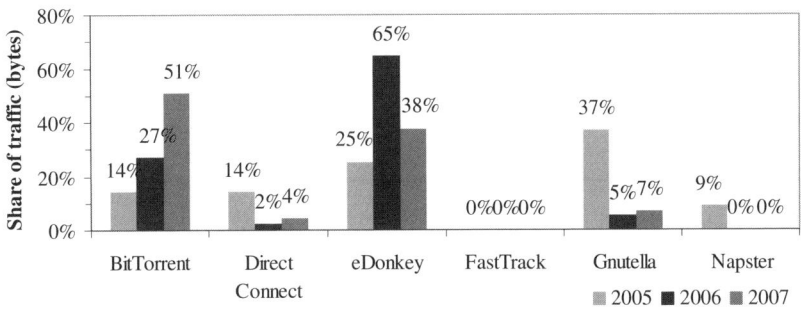

Fig. 4. Breakdown of P2P traffic by protocol

4.2   Handset Monitoring Results

In the panel study, three major categories of P2P applications are found: Voice over IP (VoIP) clients, instant messaging, and file sharing. Most of the identified P2P applications receive very little usage, typically having only a few users, which rarely activate the P2P application on a regular basis (see Table 2). Clearly MP2P applications are not popular. The average number of usage sessions per month per user is relatively high. The usage session is defined to consist of minimum 15 seconds from the activation of the application until its closure.

Table 3 presents the results of the data volume analysis in both GSM/UMTS and WLAN. MP2P applications experience very little actual usage. According to the results, people probably only experiment with MP2P applications. Only Fring exhibits wider usage, typically generating 4% of the total packet data volume of its users. On average Fring generates 1 MB per month per user. 12% of Fring traffic takes place on GSM networks (10% in EDGE and 2% in GPRS networks), and the share of UMTS networks is 16%. The share of WLAN traffic in Fring usage is 72%, possibly indicating cost avoidance behavior.

Table 2. Application usage by panelists

Application name	Used	Used every month	Used every week	Used twice per every week	Sessions per month per user
Fring	4.5%	3.5%	1.2%	1.9%	5.1
MobileMule	0.3%	0.3%	0.0%	0.2%	5.8
iSkoot	0.2%	0.0%	0.0%	0.0%	4.4
SymTorrent	0.2%	0.2%	0.0%	0.0%	6.0

Table 3. Data volume by panelists

Application name	Avg. proportion of the user's data volume	Avg. MB per month per user
Fring	4.0%	1.01
MobileMule	0.0%	0.03
SymTorrent	0.0%	0.04
iSkoot	0.0%	0.01

5   Conclusions

In the traffic trace measurement, we observe almost zero P2P file sharing traffic for handsets, but 9-18% of unidentified traffic, part of which possibly being P2P traffic. A growing trend for computer-based P2P file sharing traffic is visible in GSM/UMTS networks, BitTorrent and eDonkey being the most popular protocols. Again, a significant growing portion (10-58%) of traffic is left unidentified by our port number based identification method, potentially suggesting a noticeable increase of P2P traffic using random port numbers. Diurnal analysis of the traffic partially confirms this behavior. Only Fring, a client to several P2P-based communication services, has significant usage and data volume levels in the panel study.

Relative P2P traffic growth and changes in P2P protocol distributions in our traffic trace measurements is dependent on global P2P trends. For instance, some of the studies discussed earlier depicted a coherent trend of growth in BitTorrent usage which is also visible in our study. The other trend seen in our study common to other studies is the relative growth of unidentified traffic potentially consisting of masqueraded P2P traffic. The absolute traffic growth is probably explained by the following significant changes in the Finnish mobile market: the active marketing of flat rate data tariffs in 2006 and USB data modems in 2007 to the consumer market by major mobile operators.

In the panel study, many kinds of metrics on service usage can be derived straight from the device. The shortcomings of the panel study include the adverse selection of panelists and the amount of data available. If the dataset contained more P2P data sessions, the data traffic patterns could be studied in detail, including the distinction between WLAN and cellular P2P use.

Further research could include applying more advanced P2P traffic identification methods to traffic trace measurements, such as Domain Name Server (DNS) based analysis and detection according to statistical identification functions. The more advanced methods could facilitate the detection of other types of P2P protocols than file sharing using random port numbers, for instance Skype [26]. While applying these methods, the potential peculiarities posed by the GSM/UMTS and WLAN networks, such as longer access delays, should be taken into consideration. The panel results could be refined by having a more precise approach to MP2P application identification and by analyzing possible correlations between respondents' demographics and usage profiles.

Acknowledgments. We would like to thank Markus Peuhkuri and Timo Smura for their assistance in analyzing the data, and Heikki Kokkinen for his comments on a draft version of this paper. This research has been conducted as part of the COST IS0605 framework.

References

1. Haßlinger, G.: ISP platforms under a heavy peer-to-peer workload. In: Steinmetz, R., Wehrle, K. (eds.) Peer-to-Peer Systems and Applications. LNCS, vol. 3485, pp. 369–381. Springer, Heidelberg (2005)
2. Kivi, A.: Mobile Data Service Usage Measurements: Results 2005-2007. Technical report, TKK Helsinki University of Technology (2008)
3. Steinmetz, R., Wehrle, K. (eds.): Peer-to-Peer Systems and Applications. LNCS, vol. 3485, pp. 9–16. Springer, Heidelberg (2005)
4. Androutsellis-Theotokis, S., Spinellis, D.: A Survey of Peer-to-Peer Content Distribution Technologies. ACM Computing Surveys 36, 335–371 (2004)
5. Hofeld, T., Tutschku, K., Andersen, F.U.: Mapping of File-Sharing onto Mobile Environments: Enhancement by UMTS. In: Proc. IEEE Pervasive Computing and Communications Workshops, pp. 43–49 (2005)
6. Hossfeld, T., Tutschku, K., Andersen, F.U.: Mapping File Sharing onto Mobile Environments: Feasibility and Performance of eDonkey over GPRS. In: Proc. IEEE Wireless Communications and Networking Conference, pp. 2453–2458 (2005)
7. Hoßfeld, T., Binzenhöfer, A.: Analysis of Skype VoIP Traffic in UMTS: End-To-End QoS and QoE Measurements. Computer Networks 52, 650–666 (2008)
8. Matuszewski, M., Beijar, N., Lehtinen, J., Hyyryläinen, T.: Content Sharing in Mobile P2P Networks: Myth or Reality. Int. J. Mobile Network Design and Innovation 1, 197–207 (2006)
9. Matuszewski, M., Kokkonen, E.: Mobile P2PSIP: Peer-to-Peer SIP Communication in Mobile Communities. In: Proc. Fifth IEEE Consumer Communications & Networking Conference, pp. 1159–1165 (2008)
10. Plissonneau, L., Costeux, J.-L., Brown, P.: Analysis of peer-to-peer traffic on ADSL. In: Dovrolis, C. (ed.) PAM 2005. LNCS, vol. 3431, pp. 69–82. Springer, Heidelberg (2005)
11. Sen, S., Wang, J.: Analyzing Peer-to-Peer Traffic Across Large Networks. IEEE/ACM Transactions on Networking 12, 219–232 (2004)
12. Guha, S., Daswani, N., Jain, R.: An Experimental Study of the Skype Peer-to-Peer VoIP System. In: 5th International Workshop on Peer-to-Peer Systems (2006)
13. Karagiannis, T., Broido, A., Faloutsos, M., Claffy, K.C.: Transport Layer Identification of P2P Traffic. In: Proc. Internet Measurement Conference, pp. 121–134 (2004)

14. Ohzahata, S., Hagiwara, Y., Terada, M., Kawashima, K.: A traffic identification method and evaluations for a pure P2P application. In: Dovrolis, C. (ed.) PAM 2005. LNCS, vol. 3431, pp. 55–68. Springer, Heidelberg (2005)
15. Schmidt, S.E.G., Soysal, M.: An Intrusion Detection Based Approach for the Scalable Detection of P2P Traffic in the National Academic Network Backbone. In: Proc. Seventh IEEE International Symposium on Computer Networks, pp. 128–133 (2006)
16. Wu, C., Li, B., Zhao, S.: Characterizing Peer-to-Peer Streaming Flows. IEEE J. on Selected Areas in Communications 25, 1612–1626 (2007)
17. Bleul, H., Rathgeb, E.P.: A simple, efficient and flexible approach to measure multi-protocol peer-to-peer traffic. In: Lorenz, P., Dini, P. (eds.) ICN 2005. LNCS, vol. 3421, pp. 606–616. Springer, Heidelberg (2005)
18. Gummadi, K.P., Dunn, R.J., Saroiu, S., Gribble, S.D., Levy, H.M., Zahorjan, J.: Measurement, Modeling, and Analysis of a Peer-to-Peer File-Sharing Workload. In: Proc. ACM Symposium on Operating Systems Principles, pp. 314–329 (2003)
19. Henderson, T., Kotz, D., Abyzov, I.: The Changing Usage of a Mature Campus-Wide Wireless Network. Computer Networks 52, 2690–2712 (2008)
20. Guo, L., Chen, S., Xiao, Z., Tan, E., Ding, X., Zhang, X.: A Performance Study of Bittorrent-Like Peer-to-Peer Systems. IEEE J. on Selected Areas in Communications 25, 155–169 (2007)
21. Smith, C., Grundl, P.: Know Your Enemy: Passive Fingerprinting. Technical report, The Honeynet Project Know Your Enemy Whitepapers Series (2002)
22. Karagiannis, T., Broido, A., Brownlee, N., Claffy, K.C., Faloutsos, M.: File-Sharing in the Internet: A Characterization of P2P Traffic in the Backbone. Technical report, University of California, Riverside (2003)
23. Verkasalo, H., Hämmäinen, H.: A Handset-Based Platform for Measuring Mobile Service Usage. INFO 9, 80–96 (2007)
24. Official Statistics of Finland: Telecommunications 2006. Statistics Finland, Helsinki (2007)
25. Official Statistics of Finland: Telecommunications 2007. Statistics Finland, Helsinki (2008)
26. Baset, S.A., Schulzrinne, H.: An Analysis of the Skype Peer-to-Peer Internet Telephony Protocol. Technical report, Columbia University, New York (2004)

Monitoring the Bittorrent Monitors: A Bird's Eye View

Georgos Siganos, Josep M. Pujol, and Pablo Rodriguez

Telefonica Research
{georgos,jmps,pablorr}@tid.es

Abstract. Detecting clients with deviant behavior in the Bittorrent network is a challenging task that has not received the deserved attention. Typically, this question is seen as not 'politically' correct, since it is associated with the controversial issue of detecting agencies that monitor Bittorrent for copyright infringement. However, deviant behavior detection and its associated blacklists might prove crucial for the well being of Bittorrent as there are other deviant entities in Bittorrent besides monitors. Our goal is to provide some initial heuristics that can be used to automatically detect deviant clients. We analyze for 45 days the top 600 torrents of Pirate Bay. We show that the empirical observation of Bittorrent clients can be used to detect deviant behavior, and consequently, it is possible to automatically build dynamic blacklists.

1 Introduction

P2P file-sharing networks have brought a revolution to the way we communicate and exchange files on the Internet. The ease with which new applications and technologies can be deployed have captured the interest of the Internet community. P2P systems like Bittorrent are widely used for file transfers for they are more efficient than classical client/server architectures. The subject of distribution, however, is sometimes copyright protected material. This fact has outraged copyright holders, who try to stop or hinder the distribution of their material over P2P networks.

The tension between users and copyright protection companies has triggered an arms race, in which both sides play a hide and seek game. This arms race has lead to the current situation in which Bittorrent networks are populated by different kinds of clients. Standard clients are those whose intension is to find and share content, whereas other clients are focused in monitoring and/or preventing the activity of the standard clients. We aim to develop a methodology to automatically detect and classify the different clients that populate Bittorrent networks.

We should stress here that wide coverage accurate real-time blacklist creation is not exclusively intended to prevent detection from legit copyright protection agencies. We are not in any way condoning unlawful sharing of copyrighted material, but rather showing that Bittorrent ecosystems have more entities besides seeders, leechers and monitoring clients. We show that Bittorrent swarms contain

S.B. Moon et al. (Eds.): PAM 2009, LNCS 5448, pp. 175–184, 2009.
© Springer-Verlag Berlin Heidelberg 2009

a high density of clients belonging to Botnets, which seem to be uncorrelated with the monitoring agencies.

The research community has already start discussing ways by which Botnets can be used to attack Bittorrent [3]. Additionally, large scale Internet attacks have been reported against P2P systems, and in some cases, allegedly by established anti P2P companies [4][6]. Accurate and up-to-date methods to discriminate between *standard* and *deviant* clients will play a key role on sustaining the success of Bittorrent networks.

Paper contributions: In this paper, we provide some initial heuristics that can be used to automatically detect deviant clients.

- We analyze for 45 days the top 600 torrents of Pirate Bay. The dataset is composed of hourly snapshots that amount for over 1.8 Terabytes of data, which contain 37 million unique IPs.
- This data captures the actual empirical behavior of Bittorrent clients, both in one hour snapshots and across time.
- Access to the aggregated behavior data of Bittorrent allow us to develop heuristics that discriminate clients by their behavior.
- We show that is possible to automatically build dynamic blacklists that capture those clients that deviate from the typical behavior. This is achieved by using a fraction of the IP space when compared to the state-of-the-art blacklists.
- Additionally, we show that a significant number of deviant clients are Botnet controlled.

2   Related Work

The presence of copyright enforcement agencies monitoring peer-to-peer users is neither new nor exclusive to Bittorrent. Understanding how copyright protection agencies monitor Bittorrent has been studied by Piatek et al [7]. Their focus, however, is to characterize the methods used by enforcement agencies to detect copyright infringement. Their study provides empirical evidence that enforcement agencies rely mostly on indirect detection methods. They received more than three hundred DMCA takedown notices without uploading or downloading any file whatsoever. These DMCA takedown notices are issued by agencies that veil for the interest of copyright holders. However, in their case the *cease or face the consequence* legal threads are unfounded since they do not take part in the exchange of the copyrighted content. The authors also show that besides false positives, the indirect detection is also weak against malicious users who can easily frame virtually any network endpoint, even their own network printer.

Banerjee et al [1] observe that between 12% and 17% of the peers in Gnutella are blacklisted. Consequently, the likelihood of interacting with peers that are susceptible to monitor is 100% unless countermeasures are in place; if peers actively avoid interaction with the top-5 most prevalent IP ranges of the blacklist, the risk of detection is reduced to 1%. They also report that most of the

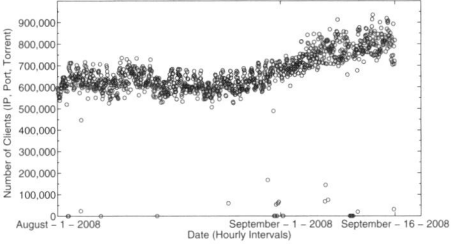

Fig. 1. Temporal evolution of the Number of clients we monitor per hour. Each dot corresponds to the number of clients captured in one hour snapshot.

encounters with blacklisted peers correspond to those peers run by copyright enforcement agencies. However, the risk reduction provided by blacklist is only effective when coverage is perfect, i.e no monitoring client goes unnoticed.

3 Measurement Details

Our analysis is based on the data collected by the Apollo project [8]. In that project, we develop an instrumented bittorrent client that can connect and exchange control messages with hundreds of thousands of other bittorrent clients within a small time frame (half to one hour).

Data collected: We monitor the 600 most popular torrents of Pirate Bay [2]. We monitor these 600 top torrents once per hour for a duration of 45 days, from August 1st, 2008 to September 16th 2008. During these 45 days period, we collected over 37 million unique IPs and 3/4 of a billion hourly instances of clients. Figure 3 depicts the number of clients analyzed on an hourly base. The number of clients per hour ranges between 500 and 900 thousand. The figure also shows the existence of seasonal effects in the Bittorrent network; the transition from August to September provides approximately 200K additional clients.

Blacklists: As a guideline to detect suspicious clients we use public available lists (blacklists) that are used to prevent establishing connections to monitoring clients. These lists are typically used in conjunction with programs like the Peer-Guardian. A repository of available blacklists can be found at [5]. We use the list provided by the Bluetack company called Level-1 which provides IP ranges for companies that either are involved in trying to stop file-sharing (e.g. MediaDefender), or produce copyrighted material (e.g. Time Warner), or are government related (e.g DoD). The list has $222,639$ ranges entries and in total it covers $796,128,149$ IPs, that is more than 47 /8.

Botnets: A serious problem in the Internet are the millions of computers that are exploited and compromised. These computers are used to run applications that are controlled not by their owners, but by the administrator of the Botnet. We want to analyze if deviant Bittorrent behavior can be partly attributed to clients

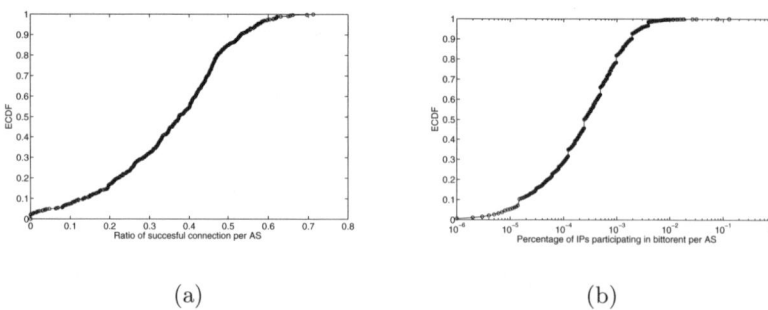

$$(a) \hspace{6cm} (b)$$

Fig. 2. (a) ECDF of the Ratio of successful connections per AS. Zero means we could not connect to any client within that AS, while one means we connected to all (b) ECDF of the Percentage of IPs of an AS that participate in bittorrent.

belonging to a Botnet. To this end, we query the Spamhaus database and check whether an IP address is part of a botnet or not. We use two lists from Spamhaus[1]. The SBL consists of IP addresses that have been verified as spam sources. On the other hand, the XBL consists of IP addresses that are known to be exploited. If an IP appears in either list, we classify it as being part of a Botnet.

4   Deviant Client Discovery

This section deals with the detection of deviant clients from the Bittorrent data previously described. We analyze the data at the level of Autonomous Systems (ASes) and at the level of prefixes and we reveal a wide variety of abnormal behavior. Further exploration of these anomalies allows us to propose a simple set of heuristics to sort out those clients that do not behave as the *standard* Bittorrent clients. These heuristics, combined with a data collection system such as Apollo [8], can lead to the automatic creation of dynamic blacklists for Bittorrent networks.

4.1   AS-Level Deviant Behavior

The first analysis is performed at the Autonomous System (AS) level in order to look for agencies big enough to have their own AS number. We focus our analysis in three different aspects and propose a heuristic to identify those ASes that are unlikely composed of *standard* Bittorrent clients. Note that the explanation is based on the data snapshot collected between 20h-21h CEST on August 1st. The observed anomalies, however, are consistent across time and the proposed heuristics do not change for different snapshots.

1. **Blocking connectivity:** Figure 2(a) shows the empirical cumulative distribution of the Ratio of successful connections per AS. We see that there are

---

[1] http://www.spamhaus.org/

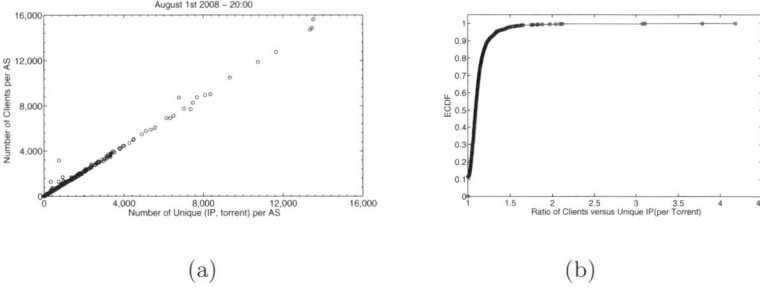

(a)   (b)

Fig. 3. (a) Number of clients versus number of IPs per AS. Ratio of over 1 means either the presence of DHCP or multiple clients per machine.(b) ECDF of the Ratio of clients versus IPs for ASes with over 100 clients.

ASes which do no accept any connection. The first heuristic rule is that ASes that do not accept incoming connections are deviant provided they have at least 100 clients. The exception is set in order to avoid the statistical effect due to a small number of clients. This rule is typically meet by 10 to 15 ASes. Using the Routing Registries the ASes can be classified into four categories: wireless, universities, ISPs with network wide firewalls, and the rest. The latter being ASes that belong to companies suspicious of carrying out monitoring activities. For instance, the last category has two ASes: AS11662 and AS33970. The first corresponds to MediaDefender, a well known company that does P2P monitoring. The second AS number corresponds to a hosting provider used by an unidentified monitoring P2P company.

2. **Percentage of active bittorrent users:** We expect that an AS used for monitoring will have a large fraction of its advertised IPs devoted to run P2P clients. Figure 2(b) shows the ECDF of the ratio between the AS's IPs that participate in the Bittorrent network and the number of advertised IPs. We find that $AS11662$ has 13% of its IPs in the Bittorrent network. The next AS has less that 1% of their IPs active. Note that using this rule we do not find AS33970, since a hosting provider has many customers that do not take part in Bittorrent.

3. **Multiple clients per IP:** Figure 3(a) depicts the number of clients in an AS versus the number of distinct (IP,Torrent) tuples. The plot clearly shows that some ASes deviate from the typical behavior. Especially, those ASes that have a small number of distinct IPs, typically less than 1,000. This can mean that either the AS is heavily using DHCP and has a small number of IPs, or that we have multiple clients running in every machine. To further explore this relation we plot, in figure 3(b), the ECDF of the ratio between the number of distinct tuples and clients for those ASes over 100 clients. As expected this plot illustrates that the most common ratio is the expected one, one client per (IP,Torrent) tuple. There exist cases in which the same IP is running a surprisingly amount of clients. We set the rule by which any AS with a ratio higher than 2 to be suspicious. For this case we identify two

ASes: the already discussed AS33970, and AS29131 that like AS33970 is a provider suspicious of hosting a monitoring P2P agency.

The heuristics can be tested against the data from a single hour snapshot. Any AS that meets at least one of these heuristic rules will have all its clients classified as deviant. In the case of the examined snapshot, we identify three ASes that meet at least one of the rules: AS11662, AS33970, AS29131.

By looking at the combination of the resources that these ASes use and the torrents they participate in it is obvious that they are doing some form of monitoring. For example, these ASes use 257, 87 and 214 unique IPs each to monitor just 2-3 movies[2].

To cross-validate our findings, we resort to comparison to the Level-1 blacklist. We find that for AS11662 and AS29131 the blacklist covers all their IPs. On the other hand, for AS33970, it only covers 55 IPs out of the 87 used. Finally, we should note two interesting observations. First, surprisingly AS33970 and AS29131 were basically monitoring the same Torrents, with the addition of two Torrents for AS29131. Second, we could not detect any monitoring activity for AS11662 and AS29131 in September. In the case of AS11662 (MediaDefender), one possible reason might be the discovery that the company had financial problems, and that the revenue from their anti-piracy services had declined considerably[3]. It is not clear, however, whether they actually stopped monitoring activities or they just found a better way to hide.

4.2   Network-Level Analysis (IP Prefixes)

After analyzing deviant behavior at the AS level, we turn into the prefix level. Suspicious clients are expected to appear more regularly than clients of standard Bittorrent users. The typical DSL customer should not be visible for more than a few hours per day. Additionally, we also expect colluded deviant clients to have a similar IP range as IPs are allocated sequentially. This way we can avoid false positives for residential clients that don't exhibit the typical behavior. Note that in the case of ISPs the same IP within a week can be assigned to different users.

Unlike the analysis on the AS level which requires only an hour snapshot, we need more than one snapshot for the prefix analysis as we check for temporal presence of clients. We chose a period of one week. Analogously to the AS level we choose the first available data range for the discussion, which in this case is the first week of August. Nonetheless, the findings and the proposed heuristics rules do not depend on this particular weekly data and are consistent across time. For the first week of August we collected $7,775,444$ unique IPs.

Figure 4(a) displays the empirical cumulative distribution of the number of times (in hours) that an IP appears over the week interval. The majority of IPs are visible for less than 24 hour per week. We label all IPs whose presence in

---

[2] Some movies have multiple torrents. For example for the movie "Wanted", they monitor 4 different Torrents.

[3] http://www.slyck.com/story1622_MediaDefender_Leak_Cost_Nearly_1_Million_Dollars

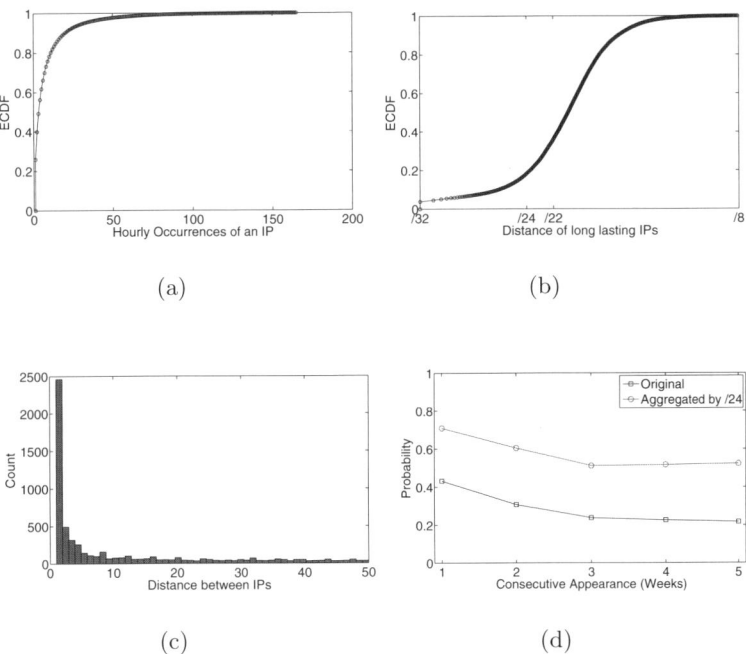

Fig. 4. ECDF of number of unique occurrences of IPs per hour. Interval August 1 - August 7 (Max 168) (b) ECDF of distance between IPs that appear at least in 50 of the hour instances. The distance is computed by sorting the IPs and find the distance between adjacent IPs. (c) Histogram of the distance between IPs (d) Probability of continuous appearance of an IP range in weeks.

the network last longer than 50 hours as possible candidates. We then proceed to analyze the distance between the IP's of the candidates. We expect deviant clients to be grouped together by IP, either sequentially or within some small distance. Figure 4(b) shows that majority of the candidates' IPs are not correlated by distance, yet some correlation exists. We group those IPs that have a distance of 6 or less to form an IP-range. IP-ranges are further grouped if they have a distance less to 100. Applying these heuristics to the first week of August yields 575 IP ranges and 6,317 IPs that classified as deviant.

Stability of ranges: Next, we check the stability of the IP ranges over periods of one week. If IP ranges are volatile, it implies that the heuristic is aggregating IP addresses by pure chance. To evaluate the stability we compute the probability of continuous appearance of an IP range throughout our measurement study. Given an IP range that appears in week i, we calculate what is the probability that it also appears in week $i + 1$. Provided that the IP range appeared in week $i + 1$, we compute the probability that it appears again in week $i + 2$ and so on. Figure 4(d) shows the probability on the original ranges in addition to the original ranges aggregated in /24s. We observe that over 20% of the original ranges are long-lasting and active during the full length of our experiment. 43%

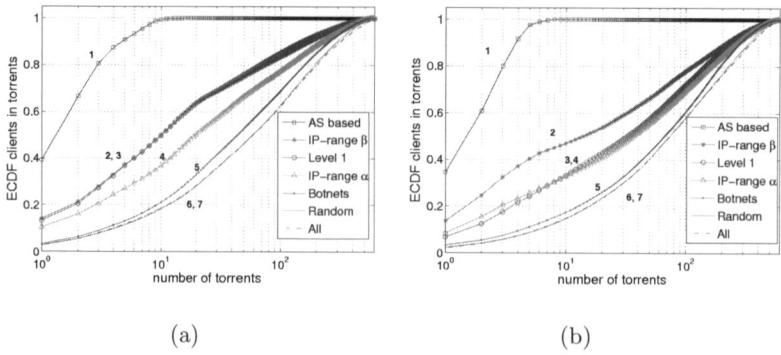

Fig. 5. (a) August 7th behavior of different blacklists (b) September 7th behavior of different blacklists. The legend order corresponds to the sequence within the plots.

of the original ranges last for at least two weeks. When aggregation takes place at the /24 level, we observe that 50% of the /24 IP ranges are active throughout our experiment and that 70% of them appear at least for 2 weeks.

5   Evaluation of Blacklist IPs

The absence of ground truth makes evaluation of blacklists a difficult problem to tackle. Accuracy of blacklists could be assessed by the coverage that different blacklists offer against receiving copyright infringement reports. However, as Piatek shows [7], DMCA takedown notices can be issued by indirect detection alone. In this case, blacklist are ineffective as they only prevent direct interaction[4].

We propose an alternative way to evaluate blacklist by observing the behavior of suspicious clients against the whole population[5]. Since most clients participating in swarms are people, we do expect that the torrents downloaded are chosen according to people's interests, and therefore, follow a pareto-like distribution. Figure 5, shows the cumulative distribution of clients by torrent for a sample of one hour for two different days: August 7th in Fig. 5(a) and September 7th in Fig 5(b). Each line corresponds to different partitions of the $667K$ clients in August and $887K$ clients in September.

Let us focus on the September data-set, since it shows more BitTorrent activity. Note that different hourly instances show quantitative variations. The

---

[4] Note, that DMCA takedown notices are only legal threats. In order to provide proof of copyright infringement, exchange of data by direct contact is necessary, which can be avoided by blacklists.

[5] We must warn the reader about a potential pitfall of this method when the criteria to partition the space of clients is a function of popularity itself. In the degenerate case, one must choose all those legitimate clients who are downloading a given movie and nothing else. This would give a popularity distribution far off the predicted by the aggregated behavior of all clients. The criteria that we use to build the partitions do not fall into this pitfall as they do not depend of the distribution itself and consequently the aggregate clients' behavior is not an artifact.

qualitative results, however, are consistent across all the samples analyzed. Taking all the $887K$ clients into consideration, *all* partition, yields the expected pareto distribution. The top-10 torrents account for 15% of the clients whereas the top-100 torrents account for 57% of clients. The remaining 500 torrents are downloaded by only 43% of clients in the BitTorrent network. The long-tail phenomenon is quite strong and shows that despite most clients are focused on *block-busters* torrents, the interest of the whole population of clients spawns over a wide range of torrents.

In order to show that the skewed distribution of clients per torrent is consistent, we pick random sets of $10K$ out of the $887K$ clients. The figure shows that the behavior of about 1% of the clients, *Random* partition, is undistinguishable from the whole population behavior depicted in the *All* partition.

The next partition to consider is based on our analysis on the the AS level (Section 4.1). This method detects $4,339$ suspicious clients, 80% of which are focused only in the top-3 torrents. Devoting so much clients in so few torrents completely ignoring the rest denotes that the clients do not follow the general trend of interest in torrents, and therefore, they are doing something else other than downloading content. We show that this behavior, yet in a smaller scale, is also observed in the other sets. Both the state-of-the-art Level-1 blacklist and our blacklist build at the level of Prefix (Section 4.2), henceforth IP-range-α, display the same *unnatural* concentration of clients in the most popular torrents at the expenses of the long-tail. Level-1 blacklist detects $7,037$ suspicious clients whereas our IP-range-α blacklist detects $18,073$. The intersection of both blacklists gives a shared $3,394$ clients. The small overlap between blacklists cannot be attributed exclusively to false positives; both blacklist have a similar distribution of clients per torrent. This finding shows that both blacklists explore different areas of the *hidden* space of deviant peers in the BitTorrent network. Different methods finding deviant clients with small overlap might indicate that the space is much larger than initially expected.

Exploring in detail individual deviant clients, we see that many are monitoring agents acting in behalf on copyright enforcement agencies. However, there are many other that are do not seem to take part in monitoring but they are detected because they are parts of a Botnet. Approximately, 39% of the deviant peers are reported as botnets by Spamhaus. We also test $10K$ IPs from the *all* set chosen at random to see if the ratio is maintained outside the suspicious client: the test yield that 2.5% of all IP's taking part in BitTorrent were botnets according to Spamhaus. Because of the high density of botnets detected by our prefix-based blacklist, we partition the *IP-range* α depending on whether the IP is found in Spamhaus or not. This yields the Botnet-free partition *IP-range* β and the only Botnet partition *botnet*, which are the two remaining partitions depicted in Figure 5.

Interestingly, the distribution of torrents by clients of *botnet* does behave very similarly to the *standard* BitTorrent clients. There is no bias in the way Botnet clients select their torrents, and consequently, monitoring is unlikely. Eliminating the botnets has the benefit of clearing out the unbiased clients leaving a more accurate blacklist of monitoring clients. The *IP-range* β partition gives $11,036$ suspicious clients, 47% of which target the top-10 movies. This is 3 times more than the expected distribution.

6 Conclusions

We evaluate the potential of using the global view of the Bittorrent network to detect deviant behavior in clients. We show that this is feasible by detecting well known monitoring clients and by discovering a large number of known Botnets, using a fraction of the IP space when compared to the state-of-the-art blacklists. We believe that the need to differentiate between a real client and a deviant client is only going to become greater in the future. Our methodology just scratches the surface of the problem. We provide some initial heuristics that can be used to automatically detect deviant clients.

References

1. Banerjee, A., Faloutsos, M., Bhuyan, L.: The p2p war: Someone is monitoring your activities. Comput. Netw. 52(6), 1272–1280 (2008)
2. The Pirate Bay. Top 100 torrents per category, http://thepiratebay.org/top
3. Defrawy, K.E., Gjoka, M., Markopoulou, A.: Bottorrent: misusing bittorrent to launch ddos attacks. In: SRUTI 2007: Proceedings of the 3rd USENIX workshop on Steps to reducing unwanted traffic on the internet, Santa Clara, CA, pp. 1–6. USENIX Association (2007)
4. Dhungel, P., Di Wu, B.S., Ross, K.W.: A measurement study of attacks on bittorrent leechers. In: Proc. IPTPS (2008)
5. iBlocklists. iblocklists, http://iblocklist.com/lists.php
6. Liang, J., Naoumov, N., Ross, K.W.: Efficient blacklisting and pollution-level estimation in P2P file-sharing systems. In: Cho, K., Jacquet, P. (eds.) AINTEC 2005. LNCS, vol. 3837, pp. 1–21. Springer, Heidelberg (2005)
7. Piatek, M., Kohno, T., Krishnamurthy, A.: Challenges and directions for monitoring p2p file sharing networks or why my printer received a DMCA takedown notice. In: Proc. of 3rd USENIX Workshop on Hot Topics in Security (HotSec 2008) (2008)
8. Siganos, G., Rodriguez, P.: APOLLO: Network transparency through a pirate's spyglass. under preparation (2009),
   http://research.tid.es/georgos/images/apollo_client.pdf

Traffic Measurements

Uncovering Artifacts of Flow Measurement Tools

Ítalo Cunha[1,2], Fernando Silveira[1,2], Ricardo Oliveira[3], Renata Teixeira[2],
and Christophe Diot[1]

[1] Thomson
[2] UPMC Paris Universitas
[3] UCLA

Abstract. This paper analyzes the performance of two implementations of
J-Flow, the flow measurement tool deployed on most Juniper routers. Our work
relies on both controlled experiments and analysis of traces collected at Abilene
and GEANT, which provide most of the flow traces used in the research community. We uncover two measurement artifacts in J-Flow traces: a periodic pattern
and measurement gaps. We investigate routers' features that trigger these artifacts
and show their impact on applications that use flow traces.

1 Introduction

Flow measurement tools summarize traffic going through routers' interfaces. These
tools aggregate packets with common characteristics (e.g., IP addresses, ports, and protocol) into flows and compute statistics such as the number of packets and bytes per
flow. Both network operators and researchers use flow statistics for tasks that range
from capacity planning and traffic matrix estimation [1] to anomaly detection [2,3].
The accuracy of these applications depends ultimately on the accuracy of statistics collected by these flow measurement tools. The first flow measurement tool was NetFlow.
It is available on most Cisco routers and has received considerable attention from the
research community [4,5,6,7]. On the other hand, Juniper's J-Flow has not been studied, even though J-Flow is used to collect statistics in Abilene and GEANT[1] (which are
among the few networks that make their data available to the research community).

In this paper, we study Juniper's J-Flow. Our analysis of flow traces from Abilene
and GEANT reveals two measurement artifacts—a *periodic pattern* and *measurement
gaps*—that appear in all their traces. We use two complementary approaches to identify
the cause and study the impact of these artifacts: (1) controlled experiments on a testbed
made of Cisco and Juniper routers running NetFlow and two different implementations
of J-Flow; and (2) analysis of real flow measurement data. Although it is possible that
NetFlow also has artifacts, we leave a specific study for future work. This paper shows
that:

- One of J-Flow's implementations exports all flows every minute. These periodic
  exports alter the duration of flows and create periodic patterns in traffic volume for
  bin sizes smaller than one minute.

---

[1] http://abilene.internet2.edu/ and http://www.geant2.net/

S.B. Moon et al. (Eds.): PAM 2009, LNCS 5448, pp. 187–196, 2009.

– There are periods when routers measure no flows. More than 1.4% of the Abilene and GEANT traces is missing. In some routers, 1% of all 5-minutes time bins lack at least 60% of their data. These gaps are correlated to routing events (i.e., routers miss data in important moments), and create dips in traffic volume.

We believe that this paper is important for all researchers working with J-Flow traces, in particular given the large data sets of publicly available traces from Abilene and GEANT. Even if the artifacts are fixed, they will be present in historical data, and it is important that the research community knows how to work with these data sets.

2 Flow Measurement Tools

J-Flow[2] and NetFlow[3] capture packets as they cross a router's interfaces and aggregate packets with common characteristics (e.g., IP addresses, ports, and protocol) into flows. They are configured per interface and support different aggregation schemes of the measured data. Flow measurement tools maintain a flow cache. Each cache entry counts the number of packets and bytes as well as keeps timestamps of the first and last packets in the flow. Flows are exported to a collector (e.g., a PC) that processes the information for visualization or stores it for later analysis.

There are three main parameters to consider when configuring flow measurements. The *inactive timeout* is the maximum amount of time flows can be in the cache without receiving packets. The *active timeout* is the maximum duration a flow can be in the cache. A flow is exported when it exceeds one of these timeouts. The *sampling rate* controls what fraction of the packets should be inspected and measured on the router's interfaces; other packets are simply ignored. Sampling is necessary because routers lack resources in order to inspect packets at very high speed links. Small timeouts cause flows to be exported more frequently, thereby increasing the accuracy of the statistics, but also increasing storage requirements at the collector. Similarly, when using a high sampling rate, the accuracy of statistics increases, at the cost of the router's CPU overhead.

We study the behavior of two different J-Flow implementations: one that shares the route processor (RP for short) with other processes, and the other that uses a dedicated measurement card (DMC).

3 Abilene and GEANT

We analyze flow statistics from Abilene and GEANT. These research backbones collect and make available different types of data sets: flow measurements, BGP updates, and IS-IS messages.

3.1 Data Description

Flow statistics are collected in all ingress links of both networks by running J-Flow in the route processor (RP). J-Flow RP is more popular than DMC because it does not

---

[2] http://www.juniper.net/products/junos/
[3] http://www.cisco.com/go/netflow/

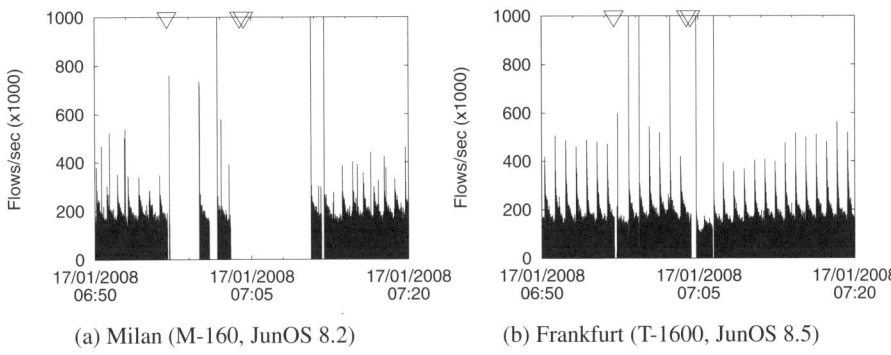

(a) Milan (M-160, JunOS 8.2)   (b) Frankfurt (T-1600, JunOS 8.5)

Fig. 1. Flow Data for GEANT

require a dedicated router port for the measurement card. Abilene currently uses Juniper T-640 routers with JunOS 8.4; and GEANT has either M-series routers with JunOS 8.2 or T-series with JunOS 8.4 or newer. J-Flow is configured with default timeouts (i.e., 15 seconds inactive and 60 seconds active); Abilene and GEANT use packet sampling probabilities of 1/100 and 1/1000, respectively. We aggregate flow statistics from all links of a router. We study data from nine routers from Abilene and 15 from GEANT for four months in four different years: 09/2004, 11/2005, 03/2007, and 01/2008.

3.2   Analysis

Fig. 1 shows the number of flows arriving at each second in two different GEANT routers. This figure shows two unexpected behaviors: a periodic pattern in the number of flows and periods during which no flow is exported.

The periodic pattern repeats every minute and appears in all Abilene and GEANT traces. This pattern is also observed in plots of the number of packets and bytes. We applied a Fourier transform on both Abilene and GEANT data, and we found significant spikes at the frequency corresponding to one minute.

The measurement gaps shown in Fig. 1 are also widespread in Abilene and GEANT data. These gaps occur often and represent between 0.05% and 2.2% of total measurement time depending on the router (i.e., between 21 minutes and 16 hours in one month). They could be explained by lost flow-export packets; however, only 0.0028% (i.e., much less then the gaps) of such packets were lost in the months considered, and even routers with no lost flow-export packets exhibit gaps. We analyze only gaps that are smaller than 10 minutes, and ignore a small number of larger gaps which might be real failures instead of measurement artifacts. While gaps are usually small, with mean gap length ranging from 2.17 to 8.8 seconds, large gaps of one to ten minutes exist, increasing the standard deviation of gap lengths up to 33 seconds.

4   Causes of Artifacts

This section investigates the causes of the periodic pattern and gaps. First, we study whether these artifacts appear in controlled experiments at a router testbed. A testbed gives us full control of the configuration of J-Flow as well as full understanding of the workload. We first describe our experiments and then analyze how the configuration of J-Flow parameters affects the measurements.

4.1 Description of Experiments

We study J-Flow with a testbed composed of two Linux machines connected to a router. One machine replays a packet trace, while the other collects the exported flows. We test a Juniper J-4350 with JunOS 8.0 and both J-Flow implementations (RP and DMC) and, for comparison, we run the same experiments with a Cisco 3825 running IOS 12.4 and NetFlow.

Each experiment proceeds as follows. We configure the router and start replaying the packet trace. Replayed packets are routed to a null interface. Exported flow records are sent to the collector. We also collect SNMP reports on packet and byte counters on the router's interfaces as well as CPU utilization. We analyze flow statistics offline after all experiments have been performed. We turn off all routing protocols and services at the routers and computers to ensure that there is no background traffic in the testbed.

We verify the accuracy of our testbed and configurations by comparing the original packet trace with that received at the collector. This analysis shows that the variance in inter-arrival times between sent and received packets is less than 2.5ms for 99% of packets. No exported flow was lost in our experiments and the SNMP data confirmed the accuracy of the measurements.

We use one synthetic and one real packet trace. The synthetic trace is designed to test whether timeouts occur as configured. It contains simultaneous flows, each with a fixed packet inter-arrival time. Inter-arrival times range from 5 to 31 seconds in steps of 0.2 seconds (i.e., 130 flows). We also replay a packet trace collected at a 100Mbit Ethernet interface between WIDE and its upstream provider[4]. We use one direction of the link in a 3-hours trace starting 14:00 on 09/Jan/07. This subset has 9402 pkts/s and 54.16 Mbits/s on average.

We denote the inactive timeout by I in seconds and the active timeout by A in minutes. Routers are configured to inspect all packets (i.e., no sampling). For the synthetic trace, the cache never overflows and all flow exports are due to timeouts only (e.g., no TCP RST/FIN packets). Results for the synthetic load are the mean over five experiments and have standard deviations under 3%.

We vary the configuration of J-Flow and NetFlow as well as the characteristics of the traces. We quantify the *precision* of each timeout as the difference between the timeout value (i.e., I and A) and measured flow durations.

4.2 Periodic Patterns

Inactive Timeout. We characterize the precision of the inactive timeout using a synthetic trace where each flow has 60 packets. We vary I between 15 and 30 seconds and fix $A = 30$ minutes. Fig. 2 shows the number of flows exported for this trace. The x-axis presents the packet inter-arrival time and the y-axis the number of flows with a

---

[4] http://mawi.wide.ad.jp/mawi/samplepoint-F/

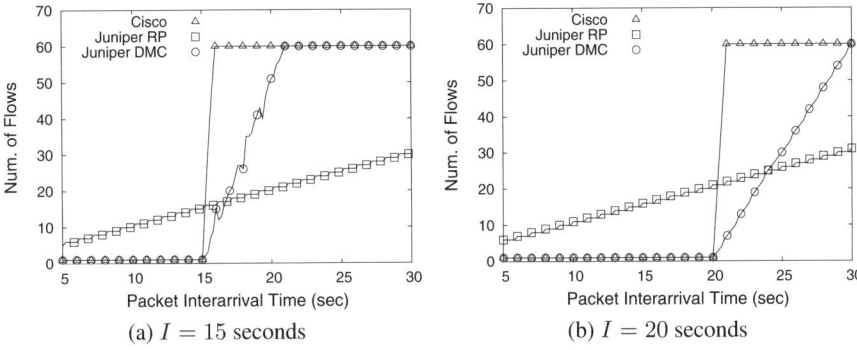

Fig. 2. Inactive Timeout Characterization

given inter-arrival time. The expected behavior of this graph is that all flows with inter-arrival times less than I should be exported in one flow record (i.e., $y = 1$), whereas flows with inter-arrival times greater than I should be exported in 60 flows (i.e., one flow per packet as after each packet, the flow created in the cache would timeout before the arrival of the next packet).

NetFlow follows the expected behavior. Its inactive timeout has a precision around one second: flows with packet inter-arrival times between I and $I + 1$ seconds are split in a variable number of flows. It is also conservative: no flows with inter-arrival times smaller than I are split. J-Flow's DMC implementation follows the same trend, but with worse precision. In addition, its precision decreases as the value of I increases. When $I = 15$ seconds, J-Flow DMC has around five seconds of precision. The precision decreases to ten seconds for $I = 20$.

Finally, a comparison of results in Figs. 2-a and 2-b shows that J-Flow's RP implementation does not consider the inactive timeout. Its behavior is exactly the same irrespective of the timeout value. In fact, the number of exported flows is proportional to the flow's duration (i.e., 60 packets × inter-arrival time) as J-Flow RP exports flows periodically every minute. We detail this behavior next.

Active Timeout. This section studies the precision of the active timeout. We run experiments with the synthetic and WIDE traces. We vary A from 15 to 30 minutes and fix $I = 15$ seconds.

Fig. 3-a shows the WIDE trace as captured by NetFlow using dots. The x-axis shows the time a flow started, and the y-axis shows the flow's duration. The WIDE trace contains many small flows. We also show as squares the results of a synthetic flow to serve as reference. This flow has one packet every five seconds and lasts for the entire experiment. Hence, given $A = 15$, we have one square every 15 minutes with flow duration of 15 minutes. In the WIDE trace, only 0.07% of flows are exported due to the active timeout. NetFlow's active timeout has a precision around two minutes (i.e., durations range from A to $A + 2$).

J-Flow DMC's behavior (not shown) is similar to the one in Fig. 3-a. However, DMC reports inaccurate start times for flows exported due to the active timeout. To show this, we plot the synthetic flow in Fig. 3-a as large circles. Instead of falling on top of

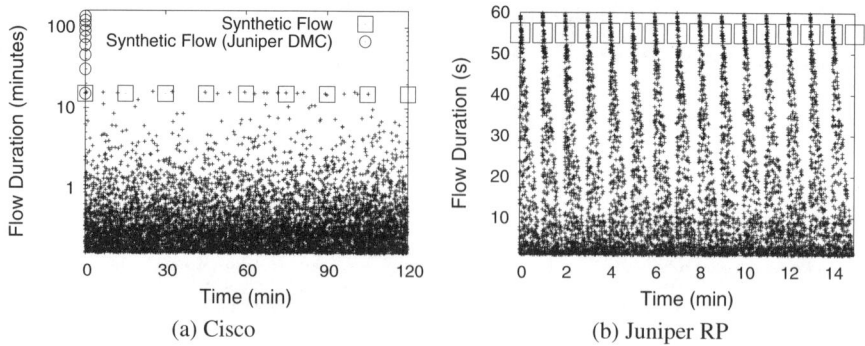

Fig. 3. Active Timeout ($A = 15$ min)

the squares, the circles are all aligned at $x = 0$, but with different durations. When DMC exports a flow due to the expiration of the active timeout it resets all the counters related to this flow, except for the start time. Hence, subsequent exports of the same flow always have the same start time and different end time, which makes the flows longer at each export.

Fig. 3-b presents the same results of Fig. 3-a when using J-Flow's RP implementation. The contrast between these two figures is striking. RP introduces a clear periodic pattern, which is similar to the pattern seen in Abilene's and GEANT's data (Fig. 1). This periodic pattern is not due to cache overflow, as it also impacts the synthetic flow. Instead of exporting the synthetic flow every 15 minutes, RP exports it every minute (as shown by the squares). Moreover, an active timeout of one minute would not create this periodicity. We conjecture that this periodic pattern is due to a flush of the flow cache every minute, when all flows are exported. This explains the sawtooth shape of Figs. 1 and 3-b. This periodic flush of the cache leads to higher memory consumption at the router [8]. For instance, the WIDE trace has an average number of active flows (i.e., memory consumption) 28% higher in J-Flow RP than in NetFlow, because the inactive timeout is ignored and short flows (specially those with one sampled packet) are kept in the cache until the next flush. Finally, frequent flushes could impose unanticipated load on the network and the collector.

Summary. Our controlled experiments show that J-Flow's RP implementation, used by both Abilene and GEANT, creates a periodic pattern in exports because it flushes the flow cache every minute. We also observe that J-Flow's DMC implementation logs inaccurate start time for flows exported due to the expiration of the active timeout. We investigate measurement gaps next.

4.3 Gaps

We did not observe measurement gaps in our testbed. This is mainly because of light load, as we could not stress the routers with only two PCs. We also lack heavy load traces obtained with NetFlow or J-Flow DMC, so we leave the evaluation of these implementations for future work. We analyze only J-Flow RP using traces from Abilene and GEANT.

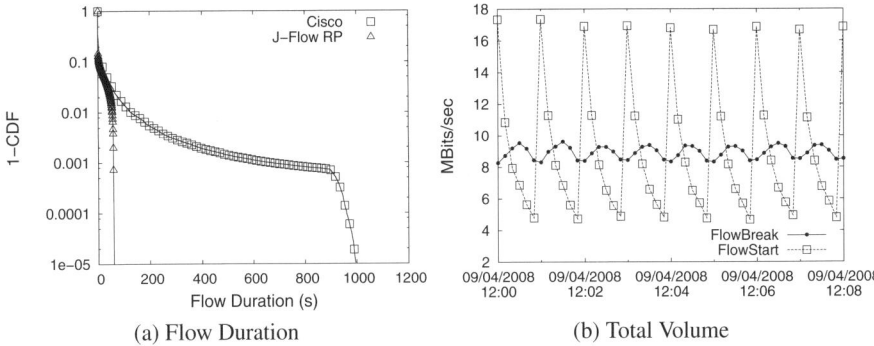

(a) Flow Duration   (b) Total Volume

Fig. 4. Impact of Cache Flushes

As J-Flow RP shares the CPU with other higher-priority routing processes, we check if routing messages are related to gaps. We use Abilene and GEANT traces to collect IS-IS updates that signal changes in the link state data base (LSDB) of each router, i.e. if an internal link fails or recovers, an IS-IS update is propagated to all routers inside the network. The instants when such messages happen are represented by the upside down triangles in Fig. 1. We observe that the starting point of the gaps in the figure coincide with IS-IS activity.

We checked with Juniper, and they have recently addressed this in Problem Report 277942. They detected that the routing daemon could lock the routing table for extended periods of time, causing the J-Flow process to block. They updated the software, making the routing daemon unlock the routing table in a timely fashion to avoid starving the J-Flow process.

5   Impact on Applications

In this section, we show how periodic cache flushes and measurement gaps can influence the result of network management tasks such as anomaly detection. We quantify this impact on the Abilene and GEANT traces, and discuss alternatives to deal with the problem.

5.1   Periodic Cache Flushes

Cache flushes have a direct impact on the distributions of flow sizes and flow durations. In J-Flow RP, the flow duration distribution will clearly have a truncation point at one minute, as shown in Fig. 4-a for the WIDE trace. We do not show the flow durations for J-Flow DMC, since this implementation does not reset the time stamp of the start of the flows after the active timeout expires. The distribution of flow sizes (not shown), which is the basis of applications such as heavy hitter detection, follows the same truncated behavior.

Cache flushes also cause periodic patterns in traffic volume over time. This behavior is only visible when traffic is binned in intervals smaller than one minute. We consider two ways of allocating bytes to bins. *FlowStart* adds all packets (or bytes) in a flow to

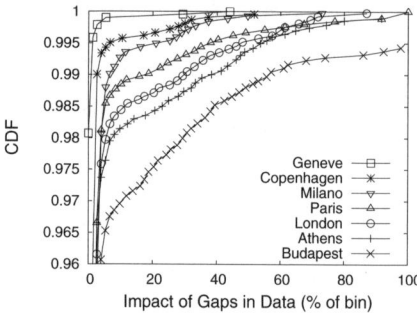

Fig. 5. Impact of Gaps on 5-min Bins

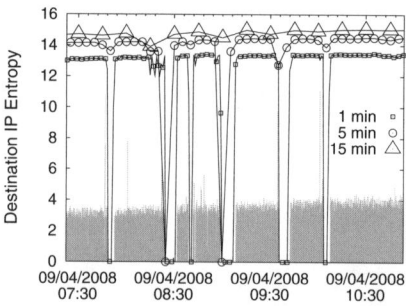

Fig. 6. Impact of Gaps on Entropy

the bin where the flow started, whereas *FlowBreak* assumes that packets in a flow are equally spaced across the duration of the flow and spreads them on the corresponding bins. Fig. 4-b shows the traffic volume for bins of ten seconds using GEANT data. FlowStart has the same periodic behavior as of Fig. 3-b, as more flows start at the beginning of each minute after the cache flush. FlowBreak is more stable, but also shows a periodic pattern. These periodic patterns are relevant to applications using flow statistics at fine-grained time granularities. The periodic behavior in Fig. 4-b is less visible if bins are larger (more aggregation) or if traffic volume is small (more variation). Bins larger than one minute are not significantly impacted by cache flushes.

Most researchers and tools already post-process flow measurements, joining exported flows with the same signature to calculate their duration. Wallerich et al. [6] presents and evaluates an effective method for this task. We point out that tools working at finer time-scales could also take advantage of the periodic exports of a flow to estimate its transfer rate at 1-minute intervals, instead of calculating an average over larger intervals.

5.2 Gaps

Large measurement gaps can significantly impact the volume of traffic in a bin. Fig. 5 shows the distribution for the amount of missing data per 5-minutes bin on January 2008 for different GEANT routers. For some routers, 1% of their bins (i.e., 90 bins over a month) have more than two minutes of missing data. These gaps would impact both anomaly detection [2,3] and traffic matrix estimation [1].

Lakhina et al. [3] proposed using the entropy of traffic feature distributions (e.g., IP addresses and ports) to find traffic anomalies. Fig. 6 shows the impact of measurement gaps on the entropy of destination IPs using different bin sizes. The behavior of other entropies (e.g., ports) are similar. The flow measurement, with gaps, is plotted in gray. Gaps have small impact on entropy because they do not impact directly the distribution of IP addresses and ports. On one hand, small bins that contain no data because of large gaps have an entropy value of zero; on the other hand, 15-minutes bins are completely oblivious to the gaps.

The gaps have two opposing effects on anomaly detection. First, a gap may be easily confused with a link failure, and detection methods can trigger alarms which are just exposing a measurement artifact instead of a traffic change. Second, the gaps are often caused by routing events, which may lead to traffic shifts that anomaly methods should be able to uncover. Since data is missing right after a routing change, it may be impossible to analyze the anomaly and quantify its impact on applications.

New routers, like Frankfurt and Geneva (Juniper T-1600 with JunOS 8.5), experience only few and short gaps, as shown in Figs. 1-b and 5, respectively. Although the impact of gaps may become less important as software and hardware are upgraded, these artifacts are present in large amounts of data collected in the past and should be taken into account by researchers and tools (e.g., check SNMP packet counters to differentiate an outage from a measurement gap).

6   Related Work

Although Cisco NetFlow has received a lot of attention [4,5,6,7], we are the first to study the correctness and accuracy of Juniper J-Flow and how its measurement artifacts impact applications. The accuracy of assuming constant throughput and packet sizes when joining flows (i.e., FlowBreak) was analyzed by Wallerich et. al. [6]; the authors found that this is a good approximation for heavy-hitters while still reasonable for small flows. Previous work have also proposed improved methodologies for flow measurement [5,8]. They are based on the idea of dynamically setting the configuration (e.g., sampling rate) depending on the available resources and the workload on the router. These works are orthogonal to ours: while a better measurement methodology is desirable, J-Flow and NetFlow are the currently deployed tools and large data sets are collected using them.

7   Conclusion

We have identified two measurement artifacts that impact most publicly available traces used by the Internet research community to study new traffic analysis and engineering techniques. At large time scales (15 minutes and above), measurement gaps generally go unnoticed. This is critical as these measurement gaps can include an anomaly as they are often correlated to major routing events that could trigger large traffic shifts. At small time scales, measurement gaps are detectable, but cache flushes create periodic patterns in traffic volume. The research community needs to be aware of these artifacts in order to take them into account when using Abilene and GEANT traces to validate their research work. These artifacts could also impact ISPs that use flow measurement tools for traffic engineering and billing.

Acknowledgements

We thank Maurizio Molina for his comments and help in understanding measurement artifacts in GEANT data. We also thank Olivier Fourmeaux and Florian Le Goff for giving us access to the LIP6 routers. We thank Jennifer Rexford, Zied Ben-Houidi, and Augustin Chaintreau for their comments on the paper.

References

1. Zhao, Q., Ge, Z., Wang, J., Xu, J.: Robust Traffic Matrix Estimation with Imperfect Information: Making use of Multiple Data Sources. SIGMETRICS Perform. Eval. Rev. 34(1), 133–144 (2006)
2. Barford, P., Kline, J., Plonka, D., Ron, A.: A Signal Analysis of Network Traffic Anomalies. In: Proc. ACM IMW, Marseille, France (November 2002)
3. Lakhina, A., Crovella, M., Diot, C.: Mining Anomalies Using Traffic Feature Distributions. In: Proc. ACM SIGCOMM, Philadelphia, PA (August 2005)
4. Sommer, R., Feldmann, A.: NetFlow: Information Loss or Win?. In: Proc. ACM IMW, Marseille, France (November 2002)
5. Estan, C., Keys, K., Moore, D., Varghese, G.: Building a Better NetFlow. In: Proc. of ACM SIGCOMM, Portland, OR (August 2004)
6. Wallerich, J., Dreger, H., Feldmann, A., Krishnamurthy, B., Willinger, W.: A Methodology for Studying Persistency Aspects of Internet Flows. SIGCOMM Comput. Commun. Rev. 35(2), 23–36 (2005)
7. Choi, B.Y., Bhattacharyya, S.: Observations on Cisco Sampled NetFlow. SIGMETRICS Perform. Eval. Rev. 33(3), 18–23 (2005)
8. Kompella, R., Estan, C.: The Power of Slicing in Internet Flow Measurement. In: Proc. ACM IMC, Berkeley, CA (October 2005)

Empirical Evaluation of Hash Functions for PacketID Generation in Sampled Multipoint Measurements

Christian Henke, Carsten Schmoll, and Tanja Zseby

Fraunhofer Institute Fokus, Berlin, Germany
{christian.henke,carsten.schmoll,tanja.zseby}@fokus.fraunhofer.de

Abstract. A broad spectrum of network measurement applications demand for multipoint measurements; e.g. one-way delay measurements or packets path tracing. A passive multipoint measurement technique is realized by generating a timestamp and a packet identifier (ID) for each packet traversing an observation point and sending this information to a common collector. The packet ID can be provided by using parts of the packet or generating a digest of the packet content. Multipoint measurements demand for high resource measurement infrastructure. Random packet selection techniques can reduce the resource consumption while still maintaining sufficient information about most metrics. Nevertheless random packet selection cannot be used for multipoint measurements, because the packets selection decisions on its path can differ. Hash-based selection is a deterministic passive multipoint measurement technique that emulates random selection and enables the correlation of a selected subset of packets at different measurement points. The selection decision is based on a hash value over invariant parts of the packet.

When hash-based selection is applied two hash values are generated - one on which the selection decision is based and a second one that is used as the packet ID. In a previous paper we already evaluated hash functions for hash-based selection. In this paper we analyze hash functions for packet ID generation. Other authors recommend the use of two different hash values for both operations - we show that in certain scenarios it is more efficient to use only one hash value.

1 Introduction

A variety of measurement applications demand for passive multipoint measurements; like delay and loss measurement or packet tracing. A passive multipoint measurement technique is realized by generating a timestamp and a unique packet ID for each packet traversing an observation point and sending this information to a common collector. The packet ID can be a combination of packet fields or a digest calculated over the packet content. In case the packet ID is unique in the measurement domain the collector can correlate the packet observations and trace the packet throughout the network. The one-way delay between two measurement points can be deduced by the timestamp difference.

S.B. Moon et al. (Eds.): PAM 2009, LNCS 5448, pp. 197–206, 2009.
© Springer-Verlag Berlin Heidelberg 2009

Because a timestamp and packet ID needs to be generated and exported for each packet, the resource consumption for passive measurements can be immense. Packet selection methods provide a solution to reduce the resource consumption. However random selection techniques are not suitable for multipoint measurements, because a random and therefore different subset of packets is selected at each measurement node, which hinders packet tracing and delay calculation.

Hash-based selection [2] is a deterministic packet selection method that selects consistent packet subsets throughout the network. Hash-based selection is realized by the following technique: parts of the packet content that are invariant between measurement nodes are extracted and used as the hash input for a hash function. The hash function with a digest length of N bits maps the hash input to a value in the range $R = [0..2^N - 1]$. The packet itself is selected if the hash value - here called selection hash - falls into a predefined selection range $S \subset R$. The selection decision for each packet along its path is the same, provided that the selected packet content (hash input), hash function and selection range are the same at the different measurement points.

When hash-based selection is applied in multipoint measurements two hash values are generated. 1) The selection hash and 2) for selected packets a packet ID which is exported to the collector. The authors of [2] [9] [11] [7] recommend to use different hash values for packet ID and selection hash. This is reasonable because 1) good hash functions for hash based selection are not necessarily good packet ID generating hash functions and 2) the packet ID collision probability increases if only one hash value is used (see Sec. 4).

We already analyzed 25 hash functions for hash-based selection in [4]. In this paper we will analyze the same hash functions on their suitability for packet ID generation based on their hash value collision probability. Hash value collisions are critical for packet ID generation because packets with the same ID cannot be distinguished and properly traced.

In contrast to other authors we propose to use only one hash value for the hash-based selection decision and packet ID. The use of only one hash value will relieve the processing capacities of the observation points. Nevertheless the use of one hash function infers more packet ID collisions, which we propose to offset by some additional selected packets. We will use a mathematical model to calculate the implications of the decision and calculate the measurement traffic increase which can compensate for additional collisions.

2   State of Art

In [2] Duffield and Grossglauser introduced the hash-based selection technique for the purpose of packet tracing. They evaluate a simple modulus hash function for hash-based selection using four different traces. Molina [7] analyzes four hash functions (CRC32, MMH, IPSX, BOB) for hash-based selection and packet ID generation. With regards to packet ID he analyzes the uniformity of hash values and the collision probability depending on the length of the hash input. In [4] we evaluated a set of 25 hash functions for hash-based selection in terms of

1) performance 2) non-linearity 3) unbiasedness and 4) representativeness. The analysis has shown that the BOB and OAAT hash function have the best overall results from this hash function collection. Nevertheless an evaluation for packet ID generation is missing. [14] compares 6 hash functions for packet ID generation (40 bytes of unprocessed fields, CRC-16, a 16-bit hash function, CRC-32, a folded 32bit MD5 and MD5-128bit) on the hash value collision probability and processing time. The unprocessed fields do not require any processing and no collisions, but imply maximum measurement traffic. The CRC-32, folded 32bit MD5 and 128 bit MD5 hash functions provide no collision in the analyzed short traces of 100,000 packets.

3   Hash Function for Packet-ID Generation

Hash value collisions are crucial for packet ID generation, because packets with colliding hash values cannot be distinguished and properly traced. In order to resolve ambiguous traffic traces the packets are discarded at the multipoint collector. The lower the collision probability of the packet ID generation hash function the fewer packets have to be discarded. The PSAMP [13] working group recommends non-cryptographical hash functions with a short digest length of 32 bits because these can be supported by low resource PSAMP devices. Cryptographical hash functions (like SHA or MD5) that are proven to have low collision probability may not be suitable for packet ID generation, because of 2 reasons: 1) the digest length is at least 128 bits long which would mean four times the measurement traffic for each packet and 2) the processing time is higher than for non-cryptographical hash functions. We will analyze a collection of 25 hash functions each with a 32 bit digest length on two criteria which are important for packet ID generation 1) uniformity of hash value distribution and 2) hash value collisions. The hash function collection is available at [1] with references to their original sources. The cryptographical SHA1 and MD5 hash function are included in the collection as well but were trimmed to 32 bits by adding each 32 bit subblock of their hash value.

3.1   Uniformity of Hash Value Distribution

Non-uniform distributed hash values pile up at specific intervals or values and hence have a higher collision probability. The chi-square goodness-of-fit test is used to analyze if the hash value distribution is significantly different from a uniform distribution. Because the amount of possible hash values R is large and only few occurrences of the same hash value are expected, the hash values are categorized in b bins that consist of $\frac{R}{b}$ values. The test statistic T is obtained by

$$T = \sum_{i=1}^{b} \frac{(n_i - \overline{n})^2}{\overline{n}} \tag{1}$$

where n_i are the observed frequencies in bin i. By dividing the number of total observed packets N by the amount of bins b, one obtains the number of expected

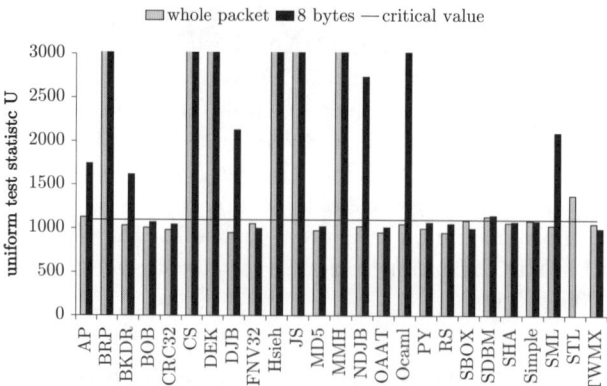

Fig. 1. Chi Square Uniformity Tests - For the test we used 5 traces, 1 million unique packets each. Hash functions whose test statistic T is higher than the critical X^2 value have a hash value distribution statistically different from the uniform distribution and presumably generate more collisions than a random generator

observations $\bar{n}$ in each bin. The hypotheses H0 that the hash value distribution is uniform is rejected if T is above the critical X^2 value with b-1 degrees of freedom with an error of α.

Experimental Setup. The uniform distribution evaluation is based on 5 traces each including N=1,000,000 unique packets, i.e. duplicate packets are removed from the trace. We removed these packets, because they distort the measurement results, packets that are identical will always have the same hash value and will be discarded for delay measurements anyways. The traces are taken from different trace groups - NZIX, Twente, FH Salzburg (accessible via [8]) and two trace groups from a european telecom operator (LEO1 and LEO2) which we already used in [5] and [4]. For the evaluation the hash input consist of the IP and Transport Header except Version/IHL, TOS, TTL, IP Checksum, Flags/Fragment Offset and IP Options. This guarantees that the hash input is unique and constant between measurement points. The hash values are categorized in $b = 1024$ bins. We are using a global error rate $\alpha = 5\%$.

Uniformity Results. The measurement results for all 25 hash functions are depicted in Fig. 1. For every hash function 5 tests are conducted based on each trace group. The problem with multiple Chi square tests is that every test inherits an error. We are not interested in individual tests that falsely reject H0, but in tests that show a truly significant deviation. A common approach when multiple statistical tests are performed is the Dunn-Sidak correction [12] which we already used in [4]. This adjusted critical X^2 value is used in Fig. 1. The hash functions that pass the uniformity test are BOB, CRC32, FNV32, MD5, OAAT, PY, RS, SBOX, SHA, SIMPLE, TWMX. These hash functions are very likely to possess low collision probability and will be further evaluated.

3.2   Mathematical Model for Collision Probability

Before we evaluate the hash function collection on their collision probability
we will describe a model to calculate the amount of expected collisions when a
random number generator is used for packet ID generation. Hash functions that
generate statistically significant more collision than a random number generator
should not be applied for packet ID generation.

Parameters of the Collision Model. Besides of the type of traffic included
in the traces the hash value collision probability depends on 1) the number of
packets N that are observed at all ingress nodes during the critical time interval
$[t_0..t_0 + t_{max}]$ and 2) the hash range size R. For accurate delay measurements
duplicate packet IDs are identified at the ingress nodes of the measurement domain (see [2]). Ingress nodes are incoming links from external routers or traffic
sources within the measured network. It is not necessary to discard all duplicate
packet IDs in the measurement domain, but only packets with equal IDs occur-
ring close together in a time interval $[t_0..t_0 + t_{max}]$, where t_0 denotes the first
occurrence of the packet ID at any ingress node and t_{max} is the maximum delay
inside the measurement domain. If two equal packet IDs occur within the critical
time interval at equal or different ingress nodes than all packet ID observations
of both packets in this time frame are discarded, because one cannot ensure to
form correct packet ID pairs. Assessing the maximum delay t_{max} is tedious and
influences the delay measurement. If the maximum delay is chosen too large the
collector 1) has to store immense amounts of packet IDs before it can purge old
reports and 2) discards more packets than are necessary. If the maximum delay
t_{max} is assessed too low, it is possible that wrong packet ID pairs are formed
and the delay measurement becomes inaccurate.

The average amount of packets N that are generated during the critical time
interval at all k ingress nodes each with a data rate B_i can be calculated using
the incoming data rate $B = \sum_{i=1}^{k} B_i$, the average packet length $\bar{l}$ and t_{max} by
$N = \frac{B \cdot t_{max}}{\bar{l}}$. As an example we assume a measurement domain with 4 backbone
ingress links each $B_i = 2500 Mbit/s$ and $\bar{l}=500$ bytes and with a maximum
domain delay of $t_{max} = 1s$. In a fully loaded case the critical interval has a size
of 2.5 million packets.

Collision Model. The amount of collision can be obtained by using a two step
model. First the probability is calculated that one specific hash value occurs m
times when N hash values are randomly generated. Second, these probabilities
for one hash value are transferred to all hash values to obtain the distribution
of the amount of unique hash values.

First Step: A random variable X is defined:

$\mathbf{X}$ = amount of draws of the specific hash value i.

Because the hash value generation is random and independent the probability
distribution of X is given by the binomial distribution $B(m|p, N)$ which can be
approximated by the poisson distribution for $N > 50$ and $p < 0.05$ and $\lambda < 9$ [10]

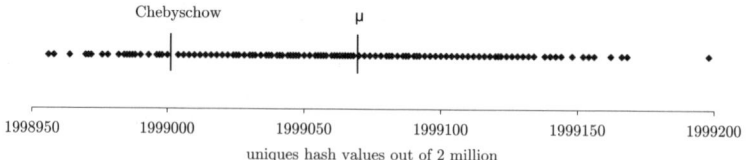

Fig. 2. In each of the 500 tests we generated 2 million random 32bit values and counted the amount of unique values. Each dot represents one test. The amount of expected unique values μ and the variance represented by the chebychev bound are shown as well. The empirical results are consistent with the mathematical model.

(which is given for our configuration) - where $p = \frac{1}{R}$ is the success probability that the specific value i is drawn in one try and $E(X) = \lambda = Np$:

$$B(m|p, N) = P(X = m) = \binom{N}{m} p^m (1 - p)^{N-m} \approx \frac{\lambda^m}{m!} e^{-\lambda} \qquad (2)$$

Second Step: We define a new random variable Y:

   Y= amount of unique hash values in the critical interval.

One approach to gain a distribution of non-colliding hash values is the following. First it is assumed that already $N-1$ hash values are generated. All hash values in the hash range R are then binomial $B(m|N-1, \frac{1}{R})$ distributed, i.e. the chance to draw a hash value that already occurred m times is approximately $P(X=m)$, because N is large. The Nth hash value is generated and it is observed how many times this hash value has been previously drawn. If the hash value has not been already drawn than a unique hash value has been generated. The probability that the hash value has not been generated before is approximately $P(X = 0)$. Moreover this conclusion can be applied for all other N-1 hash values. The success probability to draw a unique hash value is $p_y = P(X = 0)$. Because $N \ll R$ all experiments X_i can be assumed to be independent and $Y \sim B(k|N, p_y = P(X = 0))$. The expected amount of unique hash values μ and the variance σ^2 is:

$$\mu = Np = Ne^{-\frac{N}{R}} \qquad \sigma^2 = Np(1 - p) = N(e^{-\frac{N}{R}} - e^{-\frac{2N}{R}}) \qquad (3)$$

3.3   Empirical Collision Probability

First, we will use a random number generator to verify the mathematical model. Afterwards, hash functions that performed well in the uniformity tests in Sec.3.1 will be empirically analyzed on their packet ID collision probability.

Evaluation of Model using Random Generator. We used the mt19937 random number generator from the GNU scientific library [3] to obtain 32bit random values. In 500 tests the amount of unique hash values from 2 million random values are counted. For 2 million generated 32bit values the expected amount of unique hash values is (see Eq. 3) $\mu = 1999069$. The mean amount of empirical measured unique hash values in the 500 runs is $\bar{U} = 1999066$. The measurement results are depicted in Fig. 2. This figure also shows the lower bound L=1999001 for the expected amount of unique hash values in b=90% of the cases using chebyschev's inequality. 32 tests (=6.4%) show less unique hash values than the chebyschev 90% bound. The lower bound using chernoffs inequality [6] is $L_c = 1996034$ which is significantly less strict than chebychev in our setup and is therefore not further considered. The results are consistent with the mathematical model.

Empirical Evaluation of Hash Functions for Packet ID Generation on Real Traces. The empirical hash function evaluation is based on 5 traces from which all duplicate packets are removed. The traces are taken from trace group NZIX, FH Salzburg, Twente, LEO1 and LEO2 in order to ensure a variety of traffic patterns. Eleven critical measurement intervals N ranging from 1.5 to 2.5 million packets are used. As the hash input we applied the same configuration as in Sec. 3.1. The expected percentage of usable packets $\mu = E(U)$ is shown by the straight black linear line in Fig. 3. In order to represent the variance, the expected least amount of non-colliding packets in 90% of the cases is calculated using chebyshev's inequality, shown as the black dashed line. Most hash functions are very close to the expected amount of collision for all 5 traces, except the LEO2 trace. For this trace the MD5 and CRC32 hash generate 0.01 to 0.03% more colliding packet IDs than expected. Almost all other deviation can be explained by statistical variance, because they are within the 90% chebyschev bound. The BOB and RS Hash function never generated fewer unique packetIDs than predicted by the chebyschev bound.

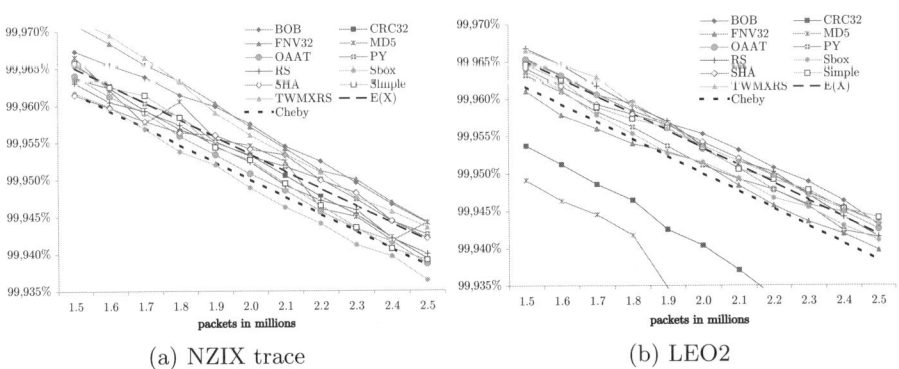

(a) NZIX trace   (b) LEO2

Fig. 3. We evaluated the hash function collection on their collision probability based on 5 different traces - only 2 depicted here. The percentage of unique hash value were compared to the mathematical model as derived with the random number generator. Most hash functions (even not cryptographically strong) are very close to the expected amount of non-colliding packet IDs. Almost all deviation are within a 90% chebyschev bound of statistical variance.

4   Packet ID and Selection Hash

When applying hash-based selection, it may or it may not be reasonable to use the same hash value which is intended for the sampling decision to use as packet ID. Using the same hash value as packet ID will relieve the measurement points processing capacities, because one does not need to calculate another hash value for the selected packets. On the other hand it will increase the collision probability of the packet IDs which can be shown by using Eq. 3. The amount of unique hash values from $s \cdot N$ packets when two hash functions U_2 or one hash function U_1 are used is:

$$U_2 = sNe^{-\frac{sN}{R}} > U_1 = sNe^{-\frac{sN}{sR}} \tag{4}$$

The increase of collision probability is the reason why [2] [7] recommend to use a different hash function for packet ID generation and packet selection. Contrary, we propose to use only one hash value for most scenarios. For example, in a scenario with a sampling fraction s close to 1, a new hash value calculation for almost all packets is not justified because there is almost no collision improvement. Instead of generating packet IDs for the sampled packets one can select additional packets in order to compensate for the additional collisions. Of course the additionally selected packets increase the measurement traffic. An example will show the trade-off. Assuming an amount of 2.5 millions packets within the critical measurement interval and a selection fraction of s=10% the amount of additional collisions $U_2 - U_1$ is 131. This means that instead of calculating an extra hash value on 250000 packets one can just select≈131 packets additionally which is an increase of measurement traffic of $\frac{C}{sN} = 0.052\%$. This value only gives an approximate figure of measurement increase because the additional selected packets have a chance to collide as well. For a more accurate value one has to look at a more practical question: How does the selection range need to be configured to gain a certain target selection fraction f_t of unique packets?

Selection Range Adjustment for One Hash Function. In the case that only one hash function is used for hash-based selection and packet ID generation, the selection range fraction of the hash range s_1 has to be configured according to Eq. 5 in order to gain a target sampling fraction f_t of usable packets.

$$f_t N = s_1 N e^{-\frac{s_1 N}{s_1 R}} \quad \rightarrow \quad s_1 = f_t e^{\frac{N}{R}} \tag{5}$$

Selection Range Adjustment for Two Hash Functions. In the case that two different hash functions are used for hash-based selection and packet ID generation there are still packet ID collisions. Therefore it is still required to adjust the selection range s_2 in order to gain a target sampling fraction t of usable packets.

$$f_t = s_2 e^{-\frac{s_2 N}{R}} \tag{6}$$

We solved Eq. 6 numerically in order to enable a comparision between both approaches.

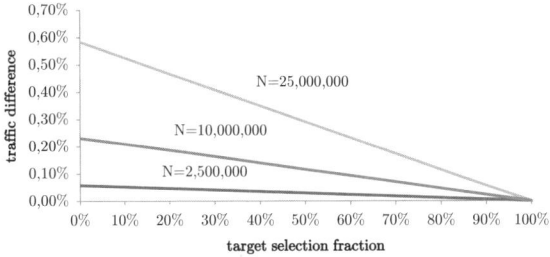

Fig. 4. Measurement Traffic Increase using 1 instead of 2 Hash Functions. When only one hash value is used for packet selection and packet ID we can relieve the measurement point, nevertheless the packet ID collision probability increases. We propose to compensate this increase by selecting some more packets, which will increase the measurement traffic. The measurement traffic increase is neglectable for large sampling fractions or small amount of packets within the critical time interval.

Difference Between One and Two Hash Function Approach. The selection range configuration differs if either one or two hash functions are used for packet ID generation and hash-based selection. The ratio $\frac{s_1 - s_2}{s_2}$ denotes the proportional difference of measurement traffic between both approaches. The increase is negligible for high target sampling ratios and small amounts of packets within the critical measurement interval N. As it is obvious from Fig. 4 the traffic increase is a linear function of the target sampling fraction. The measurement traffic increase ΔMTI can be formulated as:

$$\Delta MTI = (1 - e^{-N/R})(1 - t) \tag{7}$$

In our example with 4x2.5GB fully loaded ingress nodes and $t_{max} = 1s$ the additional measurement traffic is less than 0,06%. For scenarios with 10 / 100 as many nodes the additional measurement traffic is about 0,6% / 6% which can still justify to use only one hash function. Only in larger domains the measurement operator should use two hash functions or a longer packetIDs.

5   Conclusion

In this paper we analyzed a set of 25 hash functions on their suitability for packet ID generation. The evaluation is based on 5 different real traces. We showed that a set of 11 hash functions have comparable collision probabilities as expected from a random number generator. We recommend the BOB and RS hash function for packet ID generation because they never generated more collision than predicted by the chebychev bound and therefore they provide low and predictable collision probability.

Contrary to current approaches we analyzed the usage of only one instead of two hash functions for packet ID generation and hash-based selection. Using the mathematical model derived in Sec. 3.2 we showed for both approaches how the selection range has to be adjusted to gain a certain target sampling fraction of non-colliding and usable packet IDs. With these results we calculated the additional traffic that is required when only one hash function is used for packet ID generation and hash-based selection. For small to medium sized measurement domains it is reasonable to use only one hash function. Because the BOB hash function proofed to be under the best two functions for hash-based selection and packet ID generation, we recommend the BOB hash function in case only one hash function is used.

References

1. Hash functions,
   http://net.fokus.fraunhofer.de/research/hashfunctions.html
2. Duffield, N., Grossglauser, M.: Trajectory sampling for direct traffic observation. IEEE/ACM Trans. Netw. 9(3), 280–292 (2001)
3. Gsl - gnu scientific library, http://www.gnu.org/software/gsl/
4. Henke, C., Schmoll, C., Zseby, T.: Empirical evaluation of hash functions for multipoint measurements. SIGCOMM Comput. Commun. Rev. 38(3), 39–50 (2008)
5. Henke, C., Schmoll, C., Zseby, T.: Evaluation of header field entropy for hash-based packet selection. In: Claypool, M., Uhlig, S. (eds.) PAM 2008. LNCS, vol. 4979, pp. 82–91. Springer, Heidelberg (2008)
6. Jukna, S.: Crashkurs Mathematik für Informatiker. Teubner (2007)
7. Molina, M., Niccolini, S., Duffield, N.: Comparative experimental study of hash functions applied to packet sampling. In: ITC-19, Beijing (August 2005)
8. Traffic measurement database, http://www.ist-mome.org/
9. Niccolini, S., Molina, M., Raspall, F., Tartarelli, S.: Design and implementation of a one way delay passive measurement system. In: IEEE NOMS 2004 (2004)
10. Papoulis, A., Pillai, U.: Probability, Random Variables, and Stochastic Processes, 4th edn. McGraw-Hill, New York (2002)
11. Raspall, F., Quittek, J., Brunner, M.: Path-coupled configuration of passive measurements. In: IPS 2004, Budapest (2004)
12. Miller Jr., R.G.: Simultaneous Statistical Interference, pp. 5–8, 15–16. Springer, Heidelberg (1981)
13. Zseby, T., Molina, M., Duffield, N., Niccolini, S., Raspall, F.: Sampling and filtering techniques for ip packet selection. IETF Internet Draft
14. Zseby, T., Zander, S., Carle, G.: Evaluation of building blocks for passive one-way-delay measurements. In: PAM 2001 (2001)

On the 95-Percentile Billing Method

Xenofontas Dimitropoulos[1], Paul Hurley[2], Andreas Kind[2], and Marc Ph. Stoecklin[2]

[1] ETH Zürich
fontas@tik.ee.ethz.ch
[2] IBM Research Zürich
{pah,ank,mtc}@zurich.ibm.com

Abstract. The 95-percentile method is used widely for billing ISPs and websites. In this work, we characterize important aspects of the 95-percentile method using a large set of traffic traces. We first study how the 95-percentile depends on the aggregation window size. We observe that the computed value often follows a noisy decreasing trend along a convex curve as the window size increases. We provide theoretical justification for this dependence using the self-similar model for Internet traffic and discuss observed more complex dependencies in which the 95-percentile increases with the window size. Secondly, we quantify how variations on the window size affect the computed 95-percentile. In our experiments, we find that reasonable differences in the window size can account for an increase between 4.1% and 42.5% in the monthly bill of medium and low-volume sites. In contrast, for sites with average traffic rates above 10Mbps the fluctuation of the 95-percentile is bellow 2.9%. Next, we focus on the use of flow data in hosting environments for billing individual sites. We describe the *byte-shifting effect* introduced by flow aggregation and quantify how it can affect the computed 95-percentile. We find that in our traces it can both decrease and increase the computed 95-percentile with the largest change being a decrease of 9.3%.

1 Introduction

Transit ISPs and hosting providers monitor the traffic usage of their customers and typically charge them using the 95-percentile method. A period of a month is split into fixed size time intervals and each interval yields a sample that denotes the transferred bytes during the interval. An automated tool polls the SNMP counters of the appropriate router(s) or switch(es) and finds the transferred bytes. Then, the 95-percentile of the distribution of samples is used for billing. Often the 95-percentile is computed both on the inbound and the outbound directions and the smaller value is ignored.

In a billing cycle of 30 days, the 95-percentile method filters out 36 hours of spikes, which may include Denial of Service (DoS) attacks, flash crowds, and back-up traffic. The method essentially realizes a compromise between two objectives. The first objective is billing a customer based on its absolute traffic usage, whereas the second objective is billing based on the capacity of the provisioned links and the peak rates. If we consider the traffic rate as a continuous signal, then the first objective suggests using the average of the signal for billing, whereas the second objective suggests using the maximum. The 95-percentile is typically between the average and the maximum balancing, in this way, the two objectives.

S.B. Moon et al. (Eds.): PAM 2009, LNCS 5448, pp. 207–216, 2009.

Nevertheless, the 95-percentile is not the result of a sophisticated optimization. Certain large ISPs started using it many years ago and, over time, it became more widely-used and established. The properties and limitations (also due to operational constrains) of the method have not been systematically studied and are not well understood. For example, the window size used for computing the 95-percentile can vary between different providers. Most commonly a 5-minute interval is used, however used values range to as low as 30 seconds [1]. The effect of such variations on the 95-percentile has not been analyzed. In addition, the 95-percentile method is often applied on traffic flow data for billing, e.g., individual websites. The relationship between flow aggregation and the 95-percentile, as we discuss further in our paper, is a challenging research problem.

In this work, we characterize important aspects of the 95-percentile billing method using several traffic traces. We first analyze how the 95-percentile depends on the window size and find that it typically exhibits a noisy convex decreasing trend, although more complex inter-dependencies are also possible. Then, we make the assumption that network traffic is self-similar to provide a mathematical explanation of the observed decreasing dependence. We quantify fluctuations on the 95-percentile due to different window sizes and find that fluctuations are 1) significant, i.e., between 4.1% and 42.5% for low volume sites with average rates bellow 10 Mbps, and 2) negligible, i.e., bellow 2.9%, for high volume sites with average rates above 10 Mbps. Next, we describe how the byte-shifting effect of flow aggregation can affect the computed 95-percentile. We characterize the extent of byte-shifting and find that in our traces up to 35.3% of the total number of bytes can be shifted between neighboring windows causing a decrease on the 95-percentile by 9.3%.

The remainder of this paper is structured as follows: in the next section we describe the traffic traces we used for our experiments. In Section 3, we characterize the dependence of the 95-percentile on the size of the aggregation window. Then, in Section 4 we analyze the effect of flow aggregation on the 95-percentile billing method and provide supporting measurement results. Finally, we conclude this paper in Section 5.

2   Data Traces and Preprocessing

We used traffic traces collected with *tcpdump* [2] and *NetFlow* [3] on two distinct networks. The first network provided web hosting services to 46 websites of varying sizes. We collected unsampled NetFlow version 9 packets from the border router of the network that transferred more than 6 TBytes a day with average sending and receiving rates of 550Mbps and 100Mbps, respectively. The NetFlow trace spanned 27 days during April 2008. In addition, we used tcpdump to collect packet headers destined to or originating from an individual medium-volume website. The tcpdump trace spanned 30 days starting on the 17th of July 2007 and the average rate was 615 Kbps.

The second network was a medium-size enterprise campus network that receives transit services from a commercial and an academic ISP. In particular, we collected data from the IBM Zurich Research Laboratory campus that hosts approximately 300 employees and at least as many networked computers. For our experiments, we used a tcpdump trace collected over a period of approximately 63 continuous days, from the 2nd of March until the 5th of May 2008. The trace includes all the outgoing and incoming packet headers and the overall average traffic rate was 7.536 Mbps.

We processed the tcpdump and NetFlow traces and created a set of traffic volume time series for our experiments. We first parsed the tcpdump data collected from the campus network and computed the total (both inbound and outbound) number of bytes observed in each consecutive second. In this way, we derived a sequence of Byte counts, which we split into two time series that spanned approximately one month each.

In addition, we derived a time series for each individual website in the hosting environment. We used the NetFlow trace and associated flows with sites based on the known IP addresses of the latter. Then, we distributed the size of a flow uniformly over its time-span and derived how much each flow contributed to each one-second window. By aggregating the contributions of the flows, we constructed a baseline time series that indicates the total bytes sent and received from a site during each consecutive second. In this way, we derived a time series for each individual website. Out of the 46 sites, we ignored the time series that corresponded to the 12 lowest-volume sites that had on average a rate smaller than 1 Kbps. These sites appeared virtually unused and, therefore, we used the remaining 34 sites for our experiments. In addition, we derived one last time series from the tcpdump trace for the individual website using the procedure we outlined above.

Overall, we used 37 time series with an one-second time resolution. We call these time series baselines. To measure the effect of the aggregation window size, we aggregated the baselines using windows of varying size and computed the 95-percentile of the aggregated series that we report in our experiments.

3   95-Percentile versus Window Size

3.1   Measurements

Using the different traces, we first examine how the 95-percentile depends on the window size. In Figure 1(a), we illustrate the relationship between the 95-percentile and the window size for the traffic of the enterprise campus network. The computed 95-percentile reflects what a transit provider would charge the network. As the size of the aggregation window decreases, we observe that the 95-percentile increases. The 95-percentiles corresponding to a window size of 30 seconds increases by 5% with respect to the 95-percentile of a 300-second window.

Secondly, we examine how the 95-percentile depends on the window size in the case of a web-hosting provider charging a high-volume website. In Figure 1(b), we illustrate the corresponding plot for the website in our traces that had the highest mean traffic volume. We observe again that the 95-percentile exhibits a noisy decreasing trend as the window size increases. In this case the relative fluctuations of the 95-percentile are smaller than in the campus network. The 95-percentile increases by only 0.7% between a 300 and a 30-second window. In the set of 95-percentile values that correspond to window sizes between 30 and 400 seconds, the maximum 95-percentile increase we observed was 1.3%[1].

---

[1] Note that in Figure 1(b) we also illustrate the behavior of the 95-percentile for window sizes between 2 and 30 seconds, however, we do not use this range of values to find the maximum 95-percentile increase, as in practice such low window sizes are unlikely.

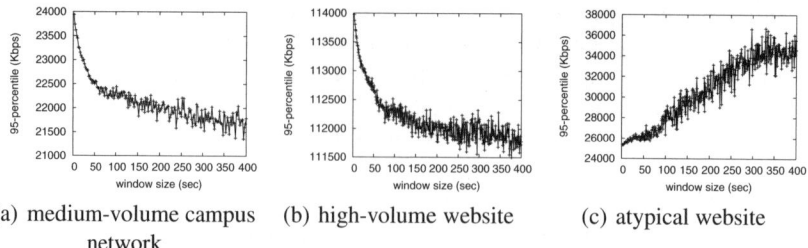

(a) medium-volume campus   (b) high-volume website   (c) atypical website
network

Fig. 1. 95-percentile versus aggregation window size

We plotted and examined the dependence of the 95-percentile on the window size for each individual site. For 23 out of the 34 sites the dependence on the window size had two characteristics in common with Figures 1(a) and 1(b): 1) the 95-percentile gradually decreases following approximately a convex curve; and 2) the 95-percentile curve is noisy often exhibiting significant fluctuations between nearby points. Among the remaining 11 sites that did not follow the identified trend, 5 exhibited a very low monthly traffic volume that on average remained bellow 50 Kbps. The last 6 medium-volume sites had irregularities in their traffic patterns, like lack of time-of-the-day or day-of-the-week effects, occasionally very high bit-rates, or long down-time periods. In Figure 1(c), we illustrate the dependence for one of the sites we identified irregularities. In this case the 95-percentile increases with the window size, which phenomenon we traced back to very high periodical traffic spikes. The spikes were close to a maximum two orders of magnitude larger than the mean traffic rate.

We further study how the fluctuations on the 95-percentile relate to the mean traffic rates of the sites. We define the *maximum fluctuation* of the 95-percentile to be the increase of the largest over the smallest value in a set of 95-percentile values. For each baseline time series, we computed aggregate time series using window sizes between 30 to 400 seconds. Then, for the aggregate time series we computed the corresponding 95-percentiles and found their maximum fluctuation. Figure 2 plots the maximum fluctuation of the sites versus their mean traffic rate. We observe that as the mean traffic rate of a site increases, the fluctuation on the computed 95-percentile decreases. For high-volume sites with mean traffic rate above 10 Mbps, the maximum fluctuation is bellow 2.9%. On the other hand, for medium and low-volume sites with mean traffic rate lower than 10 Mbps the maximum fluctuation is larger reaching up to 915% in one extreme case, but mainly varying between 4.1% and 42.5%. This observation suggests that changes in the window size can introduce notable variations in the computed 95-percentile value only on sites and networks with small and medium traffic rates. On the other hand, high-volume sites or large networks and ISPs are not significantly affected by varying the window size.

3.2   Analysis

We can model the effect of the window size as an aggregation process that takes the mean of each consecutive m samples of a traffic volume series a_s to construct an aggregate time series $a_s^{(m)}$. Taking the mean tends to decrease the volume of large samples

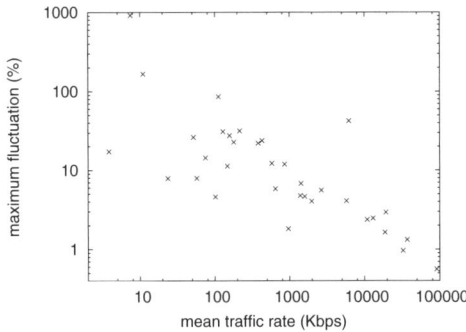

Fig. 2. Maximum fluctuation of 95-percentile versus mean traffic rate of sites

and makes the 95-percentile smaller. In theory, however, increasing the window size may also result in an increase on the 95-percentile. We illustrate this with a simple example. Assume that the baseline time series a_s corresponds to one-second time intervals and that the maximum number of aggregated samples m_u corresponds to one month. In this extreme case, the aggregated time series $a_s^{(m_u)}$ has a unique sample that is equal to the overall mean of a_s. The 95-percentile of $a_s^{(m_u)}$ is also equal to the overall mean traffic volume $\bar{a}_s$. In addition, consider that the 95-percentile of the baseline a_s can be smaller than $\bar{a}_s$, which indicates that the aggregation process will tend to increase the 95-percentile. In practice, however, real-world traffic signals typically have a mean that is smaller than their 95-percentile. The mean and 95-percentile of the baseline time series corresponding to the 1st-month trace of the campus network are 7.3 Mbps and 24.1 Mbps, respectively. As a result, increasing the window size tends to decrease the 95-percentile.

Internet traffic is known [4,5] to exhibit scaling effects. Let $X(s)$ and $X^{(m)}(s)$ denote the processes that generate a_s and $a_s^{(m)}$, respectively. If we assume that the two processes exhibit *exact self-similarity* [4] with Hurst parameter H, then their distributions are related:

$$X^{(m)}(s) \stackrel{d}{=} m^{H-1} X(s).$$

In addition, the $(1-\gamma)$-quantiles $X_{1-\gamma}(s)$ and $X_{1-\gamma}^{(m)}(s)$ of the distributions are related:

$$X_{1-\gamma}^{(m)}(s) = m^{H-1} X_{1-\gamma}(s). \tag{1}$$

For Internet traffic, the Hurst parameter H takes values between 0.5 and 1. Fixing $X_{1-\gamma}(s)$ in the last equation and setting γ to 0.05, we get that the 95-percentile of the aggregated time series decreases polynomially as the aggregation window m increases. This behavior is consistent with our observations in Figure 1 and with the remaining figures for the 23 other sites.

Besides, Figure 2 indicates that fluctuations on the 95-percentile are larger for low and medium-volume sites and smaller for high-volume sites. Higher volume sites are associated with a higher degree of statistical multiplexing. As a result, they exhibit a lower traffic burstiness than low and medium-volume sites. We speculate that this lower burstiness results in smaller fluctuations on the 95-percentile. We can use the *relative standard deviation (RSD)* of a distribution, i.e., the standard deviation divided by the mean, to quantify the burstiness of a traffic signal. The RSD is computed on the baseline time series before aggregation. In Figure 3 we illustrate the rank correlation between and RSD and the maximum 95-percentile fluctuation of the websites. We see that the points are aligned mostly along the line on the 45-degree angle, which indicates a strong correlation between the RSD and the maximum 95-percentile fluctuation. The Spearman correlation coefficient, in particular, is 0.84. This high correlation suggests that a high (low) degree of traffic burstiness results in more (fewer) 95-percentile fluctuations. Traffic burstiness can easily be quantified (using RSD) and, therefore, it can serve as indicator on how susceptible a site and a network is to 95-percentile fluctuations.

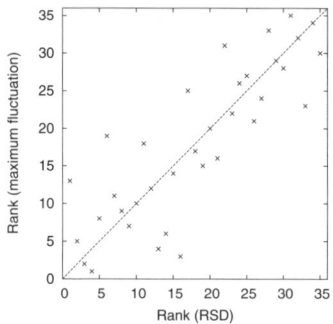

Fig. 3. Rank correlation between the traffic burstiness, i.e., relative standard deviation (RSD) of traffic rate distribution, and the maximum 95-percentile fluctuation of the different sites

In summary, our characterization and analysis of the dependence of 95-percentile on the aggregation window size yields the following important observations:

- For sites and networks of high and medium traffic rates, the 95-percentile follows a noisy decreasing trend, as the aggregation window size increases. This trend can be modeled as polynomial decrease.
- Fluctuations on the computed 95-percentile due to the size of the aggregation window are higher in low and medium-volume traffic mixes and negligible in high-volume mixes. Traffic burstiness is indicative of how susceptible the 95-percentile is to fluctuations.

4   95-Percentile from Flow Data

4.1   Measurements

NetFlow data are typically used to bill individual sites in web-hosting environments, where the traffic crossing router interfaces and incrementing SNMP counters might be destined to many different customers. NetFlow aggregates the packets of a flow and reports its duration, size, and timestamps among other attributes. Then, the 95-percentile is computed from the flow records by uniformly distributing the size of a flow over its lifes-pan and by counting the overall contribution of the flows in each aggregation window. The volume of a flow may exhibit variations, which are smoothed by aggregating the size of a flow across its duration. This "horizontal" aggregation is illustrated in Figure 4. It effectively shifts bytes between neighboring windows, which affects the estimated traffic volume for a window and may, therefore, skew the computed 95-percentile.

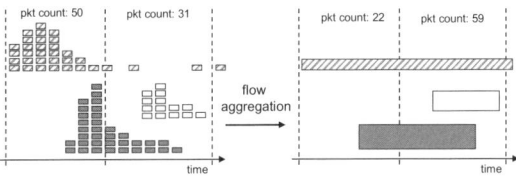

Fig. 4. Effect of flow aggregation on traffic volume observed during a time interval

To investigate the effect of flow aggregation on the 95-percentile we used the tcp-dump traces. We first computed the 95-percentiles of time series constructed from packet-level data. These time series attribute each packet to the correct window and do not suffer from the problem discussed above. Then, we grouped packets in the tcp-dump traces into flows, using the standard five-tuple flow definition, and computed the amount that each flow contributed to each window. In this case, the resulting time series were skewed due to flow aggregation.

In Figure 5, we plot the 95-percentile versus the window size for the packet and flow-based data using the tcpdump traces of the individual website and of the enterprise campus network. The plots correspond to three distinct behaviors. In Figure 5(a), we observe that the curves corresponding to using flow and packet-level data to compute the 95-percentile are almost indistinguishable, indicating negligible artifacts introduced from flow aggregation. The maximum increase of the 95-percentile in the range of window sizes between 30 and 400 seconds in only 0.42%. In Figures 5(b) and 5(c), we show the corresponding plots for the first and second month of the campus trace, respectively. Figure 5(b) exhibits small differences between the the packet and flow-based curves. For a window size of 300 seconds, the 95-percentile increases by 1%, whereas the maximum increase in the range above 30 seconds is 2.89%. Figure 5(c) demonstrates a significant decrease on the flow-based 95-percentile. This decrease is consistent through out the range of window values and has a maximum value of 9.3% at a window size of 200 seconds. At the commonly-used 300-second window size the decrease is 5.8%.

In Table 1 we illustrate the total traffic volume that was shifted due to the effect of Figure 4 between windows. We observe as expected that the volume of shifted bytes de-creases as the window size becomes larger, since fewer flows cross window boundaries. In agreement with Figures 5, the amount of shifted bytes is smaller for the website, larger for the 1st month of the campus trace and even larger for the 2nd month. In addi-tion, in Table 1 we mark the fraction of the total traffic that was shifted to a neighboring window. The traffic fraction is as large as 35.3% indicating that the effect of Figure 4 can be prevalent leading to significant distortion of a flow-based traffic signal.

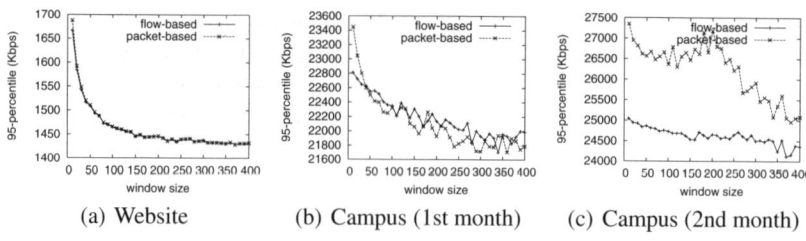

     (a) Website   (b) Campus (1st month)   (c) Campus (2nd month)

Fig. 5. 95-percentile versus window size computed from packet-level and flow-level data

Table 1. Shifted traffic between neighboring windows in our experiments

Window size (sec)	Website	Campus 1	Campus 2
	Shifted Mbytes / Shifted traffic fraction		
30	1,270 / 2.4%	201,431 / 30.2%	248,892 / 35.3%
100	134 / 0.8%	48,705 / 24.4%	63,748 / 30.1%
200	38 / 0.5%	20,112 / 20.1%	27,583 / 26.1%
300	19 / 0.3%	11,820 / 17.7%	16,714 / 23.7%

4.2 Analysis

We further investigated the traces to understand the reasons leading to the three distinct behaviors illustrated in the above figures. Figures 5(b) and 5(c) correspond to two consecutive months in the same network and the significant difference in the second month worthed further examination. Our investigation revealed during a period of a week in the second month hourly large traffic spikes that persisted even during low-volume periods, like nights and weekends. Figure 6 compares the traffic patterns in the week with the spikes with another week in the first month of the trace. The periodic (hourly) spikes

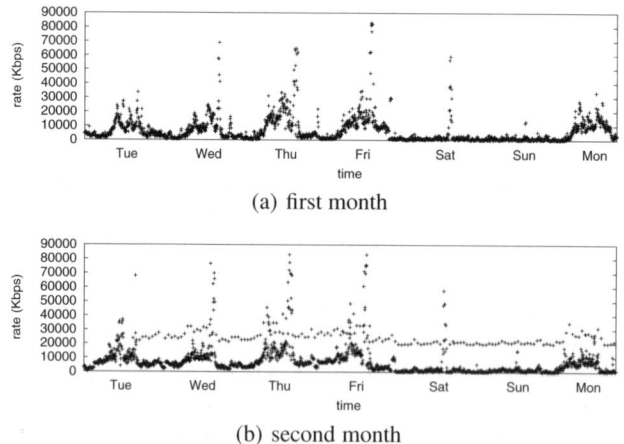

(a) first month

(b) second month

Fig. 6. Weekly traffic rate variations in the first and second month of the enterprise campus network

shown in Figure 6(b) are ranked in the top 5% of the monthly samples. Without flow aggregation they increase the 95-percentile, however, with flow aggregation the spikes are smoothed and therefore do not significantly affect the 95-percentile.

On the other hand, the high agreement between the packet and flow-based curves in Figure 5(a) results from the distinct properties of the trace. In particular, the website trace includes mainly short-lived http flows, which are less susceptible to the effect of Figure 4 than the more diverse set of traffic flows encountered in the campus network. For this reason, we only observe very few bytes shifted between windows in Table 1.

In summary, our analysis illustrates that 'horizontal" aggregation due to flow accounting skews the computed 95-percentile. The (lack of) decrease/increase on the 95-percentile depends significantly on the individual characteristics of the examined traffic traces and ranges from a 2.89% increase to a 9.3% decrease.

5   Discussion and Conclusions

In this work, we used a large set of data to study the widely-used 95-percentile billing method. We make a number of observations: 1) for medium and high-volume traffic rates the 95-percentile typically decreases as the aggregation window size increases; 2) more complex relations between the window size and the 95-percentile are possible; 3) the observed 95-percentile fluctuations were significant only for medium and small-volume traffic rates and rather negligible for high-volume sites; 4) flow aggregation can skew the computed 95-percentile value causing in our data a decrease up to 9.3%. Besides, we used certain properties of Internet traffic to justify our observations and provide a mathematical basis.

A natural question to ask is how to compute the 95-percentile correctly? One could make the assumption that the traffic rate is a continuous signal and could try to find the 95-percentile of the continuous signal. In this case, the 95-percentile would be well-defined. In our early work, we experimented with trying to find an "ideal" 95-percentile of an assumed underlying continuous signal. However, it turned out that the Fourier spectrum of network traffic has many high frequency components, which by the well-known standard Shannon sampling theorem would require sampling network traffic with a very high frequency or equivalently aggregating network traffic using a very small window. Such a "correct" 95-percentile, therefore, would be impossible to compute in practice due to the high measurement and instrumentation overhead it would require.

One take-away of our work is that providers should *all* use a fixed, ideally standardized, window size to charge their customers, in order to enable a fare comparison between different billing rates. This might be already happening to a certain extent, as the 5-minute window size is popular. However, not everybody uses the same window size and more importantly the over-charging consequences of varying the window size were not known, to the best of our knowledge, before our work.

A second take-away of our work is that "horizontal" aggregation introduced by flow accounting can skew the number of bytes during a window interval, which, in turn, can bias the computation of the 95-percentile. This observation is significant, as flow technologies are widespread for billing. A possible simple solution to this problem is to have synchronized routers/switches that at fixed timestamps, e.g., at 16:00, 16:05, ..., expire flows. This aligns flow durations within the aggregation window intervals and, therefore, the described byte-shifting effect is avoided. This approach, if not implemented intelligently, however, could lead to flow export synchronization problems.

Summing up, in the future we would like to understand better the properties of network traffic that affect the computed 95-percentile from flow data. An important challenge is collecting long, i.e., ideally at least one month long, tcpdump traces from different networks. In addition, a model of the byte-shifting process described in Section 4 could have several applications, like in predicting 95-percentile changes or in reconstructing accurate traffic time series from flow data.

References

1. Webhostingtalk Forum: 95th percentile billing polling interval (2008) (last accessed: 09/23/2008), http://www.webhostingtalk.com/showthread.php?t=579063
2. The Tcpdump team: tcpdump, http://www.tcpdump.org/
3. Cisco: IOS NetFlow, http://www.cisco.com/en/US/products/ps6601/products_ios_protocol_group_home.html
4. Leland, W.E., Taqqu, M.S., Willinger, W., Wilson, D.V.: On the self-similar nature of ethernet traffic. SIGCOMM Comput. Commun. Rev. 23(4), 183–193 (1993)
5. Paxson, V., Floyd, S.: Wide-area traffic: the failure of poisson modeling. In: SIGCOMM 1994: Proceedings of the conference on Communications architectures, protocols and applications, pp. 257–268. ACM, New York (1994)

Measurements of Anomalous
and Unwanted Traffic

Dynamics of Online Scam Hosting Infrastructure

Maria Konte[1], Nick Feamster[1], and Jaeyeon Jung[2]

[1] Georgia Institute of Technology
[2] Intel Research
{mkonte,feamster}@cc.gatech.edu, jaeyeon.jung@intel.com

Abstract. This paper studies the dynamics of scam hosting infrastructure, with an emphasis on the role of fast-flux service networks. By monitoring changes in DNS records of over 350 distinct spam-advertised domains collected from URLs in 115,000 spam emails received at a large spam sinkhole, we measure the rates and locations of remapping DNS records, and the rates at which "fresh" IP addresses are used. We find that, unlike the short-lived nature of the scams themselves, the infrastructure that hosts these scams has relatively persistent features that may ultimately assist detection.

1 Introduction

Online scam hosting infrastructure is critical to spam's profit cycle; victims must contact point-of-sale Web sites, which must be both highly available and dynamic enough to evade detection and blocking. Until recently, many sites for a scam were hosted by a single IP address for a considerable amount of time (i.e., up to a week) [2]. However, simple countermeasures, such as blocking the IP address called for more sophisticated techniques. For example, the past year has seen the rise of "fast-flux service networks" [5], which allow the sites that host online scams to change rapidly.

This paper studies the dynamics of the Internet infrastructure that hosts point-of-sale sites for email scam campaigns. We focus on how fast-flux service networks are used to host these online scams. Beyond offering a better understanding of the characteristics of the infrastructure, our study discovers invariant features of the infrastructure that may ultimately help identify scams and the spam messages that advertise them faster than existing methods.

We study the scam sites that were hosted by 384 domains as part of 21 scam campaigns in over 115,000 emails collected over the course of a month at a large spam sinkhole. This paper studies two aspects of the dynamics:

- *What are the rates and extent of change?* We examine the rates at which scam infrastructures, via the use of fast-flux service networks, redirect clients to different authoritative name servers (either by changing the authoritative nameserver's name or IP address), or to different Web sites entirely. We find that, while the scam sites' DNS TTL values do not differ significantly from other sites that perform DNS-based load balancing, the rates of change (1) differ from legitimate load balancing activities; and (2) differ across individual scam campaigns.

S.B. Moon et al. (Eds.): PAM 2009, LNCS 5448, pp. 219–228, 2009.

Table 1. DNS lookup results for the domain `pathsouth.com` (responding authoritative name-server was 218.236.53.11): The IP addresses in bold highlight changes between the two lookups taken six minutes apart. For the full list of IPs for this domain, see our technical report [7].

Time: 20:51:52 (GMT)					
A records		TTL	NS records	TTL	IPs of NS records	TTL
77.178.224.156, 79.120.37.38,		300	ns0.nameedns.com,	172800	218.236.53.11,	172800
79.120.72.0, 85.216.198.225,			ns0.nameedns1.com,		89.29.35.218,	
87.228.106.92, 89.20.146.249,			ns0.renewwdns.com,		78.107.123.140,	
213.141.146.83, 220.208.7.115			ns0.renewwdns1.com		79.120.86.168	
Time: 20:57:49 (GMT)					
A records		TTL	NS records	TTL	IPs of NS records	TTL
61.105.185.90, 69.228.33.128,		300	ns0.nameedns.com,	172800	218.236.53.11,	172800
79.120.37.38, 87.228.106.92,			ns0.nameedns1.com,		89.29.35.218,	
89.20.146.249, **89.20.159.178,**			ns0.renewwdns.com,		78.107.123.140,	
89.29.35.218, 91.122.121.88,			ns0.renewwdns1.com		**213.248.28.235**	

- *How are dynamics implemented?* We study the mechanics by which scam hosting infrastructures change the Web servers to which clients are redirected. We determine the location of change by monitoring any changes of (1) the authoritative nameservers for the domains that clients resolve (the NS record, or the IP address associated with an NS record) or of (2) the mapping of the domain name to the IP address itself (the A record for the name). We analyze both on the basis of individual spam-advertised domains and campaigns that are formed after domain clustering. We find that behavior differs by campaign, but that many scam campaigns redirect clients by changing *all three* types of mappings, whereas most legitimate load-balancing activities only involve changes to A records. We also study the infrastructures in terms of the geographical and topological locations of scam hosts and the country in which the domains were registered.

Background. Fast-flux is a DNS-based method that cybercriminals use to organize, sustain, and protect their service infrastructures such as illegal Web hosting and spamming. Somewhat similar to a technique used by content distribution networks (CDNs) such as Akamai, a fast-flux domain is served by many distributed machines, and short time-to-live (TTL) values allow a controller to quickly change the mapping between a domain name and its A records, its NS records, or the IP addresses of its NS records) [13]. Cybercriminals can rotate through compromised hosts, which renders traditional blacklisting largely ineffective. We show an example of a fast-flux domain, called `pathsouth.com`, that we monitored on January 20, 2008 (Table 1).

Related Work. The operation of fast-flux service networks and the use of these platforms to send spam was first described in detail by the Honeynet Project [13]. Compared to other studies of fast-flux networks [15, 4, 8], we focus on fast-flux networks as they relate to hosting online scams. This paper is the largest such study (it is an order of magnitude larger than the previous study [4]), and it is the first to (1) study the location (within the DNS hierarchy) of dynamics, (2) the behavior of hosting infrastructure across campaigns. We examine scam hosting infrastructure using both spam trap data and content-based scam campaign clustering; we draw on previous studies that analyzed spam trap data [6, 11, 14] or performed content-based analysis [2, 4, 9], albeit for different purposes. Previous work has used passive DNS monitoring to study the dynamics of botnets [10, 3], some of which are now believed to used to host fast-flux networks.

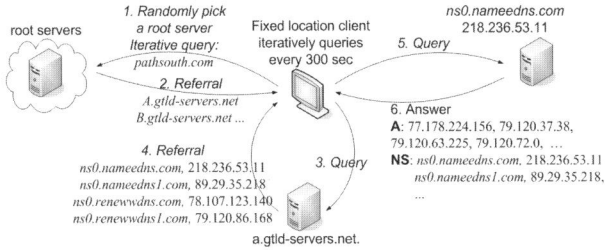

Fig. 1. Data collection diagram. The resolver records all DNS mappings at each level of DNS resolution. Here, we feature the same domain as in Table 1.

The rest of this paper is organized as follows. Section 2 describes our data and its limitations. Section 3 describes the dynamics of scam infrastructure and fast-flux service networks that we observed hosting 21 different spam campaigns over the course of a month. Section 4 refers to topological and geographic location of the observed infrastructures. Section 5 concludes with a summary and discussion of future work.

2   Data Collection

Our data collection and processing involves three steps: (1) passive collection of spam data; (2) active DNS monitoring of domains for scam sites contained in those spam messages; (3) clustering of spam and DNS data by scam campaign. This section describes these methods in detail, our use of popular Web sites as a baseline for comparison, and the limitations of our dataset.

We collected 3,360 distinct domain names that appeared at spam email messages from a large spam sink hole from October 1, 2007 to December 31, 2007. We used a simple URL pattern matcher to extract URLs from the message body. Next, we implemented an iterative resolver (at a fixed location) to resolve every domain name from this set once every five minutes. Figure 1 illustrates the process by which our resolver recorded DNS mappings at each level of DNS resolution, which allows us to monitor fast-flux networks for DNS changes at *three distinct locations in the hierarchy*: (1) the A record; (2) the NS record; and (3) the IP addresses corresponding to the names in the NS record. To avoid possible caching effects, the resolver randomly selected a DNS root server at each query. The iterative resolver recorded responses received at every level of the DNS hierarchy, including all referrals and answers.

Due to the sheer number of DNS lookups required to monitor the domains arriving at the spam trap, the resolver proceeded through the list of domains sequentially: We began by resolving the first 120 domains received at the spam trap each day. Every day the resolver added 120 new domains to the list. After each domain had been resolved continuously for three weeks, we removed the domain from the list. The resolver operated from January 14, 2008 to February 9, 2008. We picked the domains and the campaigns they mapped to, by restricting our analysis to the domains that had reachable Web sites and for which we had observed at least one change in any DNS record. To compare the dynamics those of "legitimate" domains, we used the same iterative resolution process to study the dynamics of the 500 most popular Web server domains, according to Alexa [1].

Table 2. Statistics for the top three scam campaigns compared to Alexa domains. Campaigns are sorted by the total number of IP addresses returned from A records.

Campaign	Spam emails	Spam advertising IPs	Campaign domains	Fluxing Domains	IPs of A rec	IPs of NS rec	IPs of both A+NS rec
Pharmacy-A	18459	11670	149	149	9448	2340	9705
Watch-A	40681	30411	34	30	1516	225	1572
Watch-B	454	427	43	19	1204	219	1267
All campaigns	115198	77030	465	384	9521	2421	9821
Alexa data set				500	1048	852	1877

Clustering spam by scam campaign. To cluster the spam messages into *scam campaigns*, we retrieved content from the URLs in the email messages and cluster emails whose URLs retrieve common content. We manually went through snapshot images and cluster URLs if the site is selling the same products under the same brand name using a similar page layout. In the case of slow response or when only a few small non-image files are received, we checked whether the downloaded file names of each URL is a subset of those of already identified campaign.

All 21 campaigns exhibited fluxing behavior in their DNS records to some extent during the measurement period. Table 2 shows the summary data for the three campaigns that used the most hosting servers. We denote each campaign with a *category-ID*, which we assigned based on the products offered on the Web site. The first two columns show the number of total spam emails containing the fluxing domains that we received at our spam trap and the total number of sender IPs of those spam emails. The third column is the total number of domains for that campaign, and the fourth column is the number of domain names that we found changing ("fluxing domains"). The last three columns show (1) the distinct number of IPs returned as A records of domains ($IP_{domains}$); (2) the number of IPs returned as A records of name servers ($IP_{nameservers}$); and (3) the total distinct number of IPs from the combined sets ($IP_{domains} \cup IP_{nameservers}$).

The top campaign is Pharmacy-A, one of the Canadian Pharmacy scam campaigns [12]. The campaign used at least of 9,448 distinct IP addresses as hosting servers (or front end proxies of them) for 149 domains over one month. The next two followers are Watch-A (Exquisite Replica) [12] and Watch-B (Diamond Replicas) [12], both of which offer replica watches. For these campaigns, the average number of A records associated with a single domain name is over 50, demonstrating a lot of activity in moving scam sites. We also witnessed multiple domains that shared a few hosting servers.

Registrars. To determine the registrar responsible for each of the 384 scam hosting domains, we performed jwhois queries on May 7, 2008 for each domain. Table 3 shows that about 70% of these domains are still marked as active and registered with just eight registrars in China, India, and US. Among these, the three registrars in China are responsible for 257 domains (66% of the total or 95% of the active ones). Our data collection was done before February 2008, so all domains were registered before that time. All 384 domains were all active after four months, and 2% of the domains had been active for over 7 months. Interestingly, over 40% of these domains were registered in January 2008, just before the scams themselves were hosted; thus, a newly registered domain might also ultimately serve as a useful indicator for detecting scam hosting.

Table 3. Registrars of the 384 scam domains as of May 7, 2008

Registrar	Country	Domains	Registrar	Country	Domains
dns.com.cn	China	180 (46.9%)	leadnetworks.com	India	3 (0.8%)
paycenter.com.cn	China	65 (16.9%)	coolhandle.com	US	2 (0.5%)
todaynic.com	China	12 (3.1%)	webair.com	US	1 (0.3%)
signdomains.com	India	7 (1.8%)	stargateinc.com	US	1 (0.3%)

total active domains: 271 (70.6%)

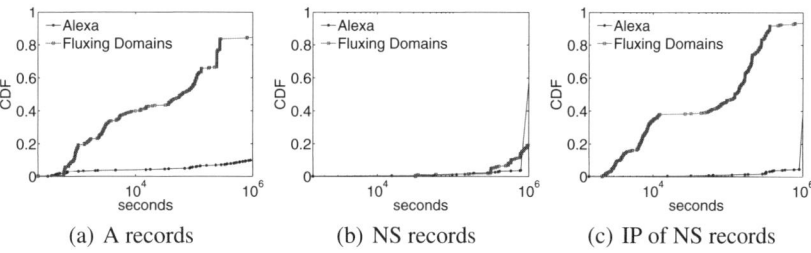

(a) A records   (b) NS records   (c) IP of NS records

Fig. 2. Distribution of the average time between changes of A, NS, and IP of NS records

Limitations. Our data is derived from spam collected at a single spam trap, which receives a relatively high number of spam messages (6,247,937 messages from October 2007 through February 2008) but may still reflect some bias in the spam it receives. Because we are primarily looking to analyze the dynamics of widespread campaigns (i.e., domains that are likely visible at many traps), this limitation should not greatly affect our results. The main limitation is that our data may not contain all domains for a particular scam. Some of our measurements occurred months after the spam was received, but our results suggest that the dynamics of these domains remain relatively consistent over the month that we monitored them.

3   Dynamics

We studied three aspects of dynamics: (1) the rate at which DNS records change at each level of the hierarchy; (2) the rate at which scam hosting infrastructure accumulates new IP addresses (both overall and by campaign); and (3) the location in the DNS hierarchy where changes take pace. To understand the nature of these features with respect to "legitimate" load balancing behavior, we also analyzed the same set of features for 500 popular sites listed by Alexa [1] as a baseline.

Rate of Change. We studied the rates at which domains for online scams changed DNS record mappings and the corresponding TTL values for these records. We compared with the TTLs for domains listed by Alexa. (Our technical report includes the TTL distribution graphs [7].) The distribution of A record TTLs shows that scam sites have slightly shorter TTL values than popular Web sites; however, both classes of Web sites have A records with a wide range TTL values. Even more surprisingly, almost all scam domains we analyzed had TTL values for NS records of longer than a day. These results make sense: many clients visiting scam sites will visit a particular domain infrequently, and only a

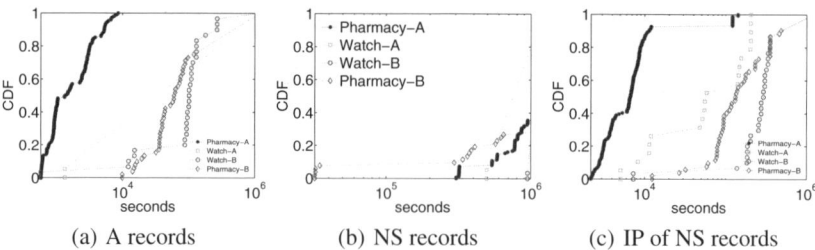

(a) A records   (b) NS records   (c) IP of NS records

Fig. 3. Cumulative distributions of the average time between changes of A, NS, and IP of NS records for Pharmacy-A, Watch-A, Watch-B, and Pharmacy-B

small number of times, so the TTL value is less important than the rate at which the mapping itself is changing (i.e., for *new* clients that attempt to resolve the domain).

We grouped the responses according to the authoritative server that provided them to account for possible load balancing. We then performed pairwise comparisons across each group of records. In the case of A and NS-record responses, we considered a response to be a change if at least one new record appears or if the number of records returned has otherwise changed since the last response; we did not consider reordering the records as a change. In the case of IP addresses of NS records, we considered the response to be a change if either NS names appear with different IPs or a new NS name appears. We discovered the following two characteristics, both of which might ultimately help automatically detect scams:

- *Scam domains change on shorter time intervals than their TTL values.* Figure 2 shows the cumulative distribution of average time between changes for each domain across all 21 scam campaigns; each point on the line represents the average time between changes for a particular domain that we monitored. The distribution shows that scam domains change hosting servers (A records) and name servers (IP addresses of NS records) more frequently than popular Web servers do, and also much more frequent than TTL values of the records.
- *Domains in the same campaign exhibit similar rates of change.* We also analyzed the rate of change of DNS records after clustering the scam domains according to campaign. Figure 3 shows these results for the top 4 campaigns (ranked by the number of distinct IPs returned in A records for domains hosting the campaigns). The results are striking: different scam campaigns rotate DNS record mappings at distinct rates, and the rates at which DNS records for a particular campaign are remapped are similar across all domains for a particular scam.

Rate of Accumulation. We measure the rate at which the network grows over time. In practice, our measurement is limited by the rate at which a domain updates its DNS records and what we present in this section is the rate at which a previously unseen host becomes an active hosting server (A records of a domain) or a name server (IP addresses of names returned by NS records).

Using a method similar to the one used by Holz *et al.* [4], we determined the rate of growth by repeatedly resolving each domain and assigning an increasing sequential ID to each previously unseen IP address. Holz *et al.* performed this analysis for A records

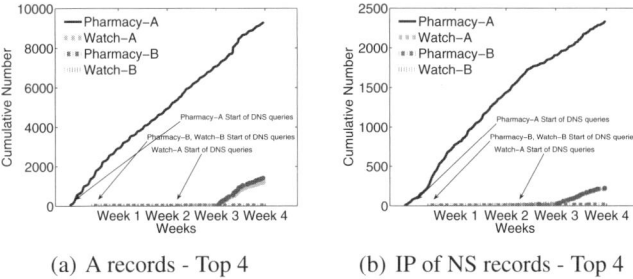

(a) A records - Top 4   (b) IP of NS records - Top 4

Fig. 4. Cumulative number of distinct IPs for the A records and IP addresses of NS records for the top 4 campaigns across the four weeks of data collection

Table 4. Location of change for the top five campaigns, sorted by the total number of distinct IPs of A records

Campaign	Domains	Location of change						
		A	[NS IP]	NS	A+[NS IP]	A+NS	NS+[NS IP]	A+NS+[NS IP]
Pharmacy-A	149	-	-	-	77	-	-	72
Watch-A	30	4	1	-	24	-	-	1
Watch-B	19	-	18	-	-	-	-	1
Pharmacy-B	52	5	13	-	19	-	-	15
Casino-A	6	-	1	-	5	-	-	-
Total	384	18	52	3	219	1	-	91
Alexa	500	37	5	15	4	1	1	-

of domains without regard to campaign; we performed this analysis for A records and IP addresses of NS records, both at the level of individual domains and at the level of campaigns:

- *Rates of accumulation differ across campaigns.* Figures 4(a) and 4(b) show the total number of distinct IPs for each scam domain (the y-value of the end of each line) over the four weeks of data collection (all iterations, 300 seconds apart from each other) and how fast each campaign accumulated new hosts (slope), for the IP addresses of A records and NS records, respectively. A steeper slope indicates more rapid accumulation of new IP addresses for that campaign.
- *Some domains only begin accumulating IP addresses after some period of dormancy.* Some domains appear to exhaust available hosts for a while (days to weeks) before accumulating new IP addresses. We examined two campaigns that exhibited rapid accumulation of IP addresses after some dormancy. Interestingly, only one domain from each campaign begins accumulating IP addresses. These two campaigns shared exactly the same set of NS names. In addition to accumulation, we also saw attrition: 10% of scam domains became unreachable in the while we were monitoring them. These domains may have been blacklisted and removed by registrars or the scammers.

Location of Change in DNS Hierarchy. We observed many scam domains with NS records or IP addresses of NS records that change rapidly, often in conjunction with other records: Campaigns change DNS record mappings at different levels of the DNS hierarchy. Table 4 shows the type of change for the top five campaign. In contrast to

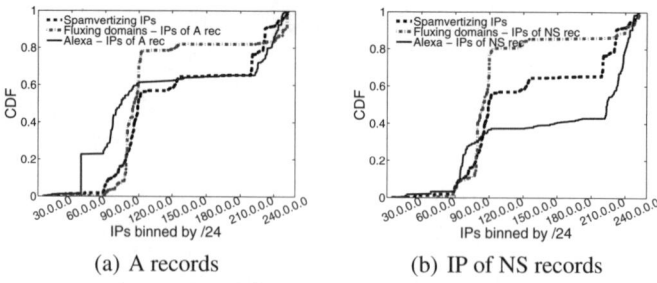

Fig. 5. Distribution of the IPs of A rec of authoritative servers spam senders

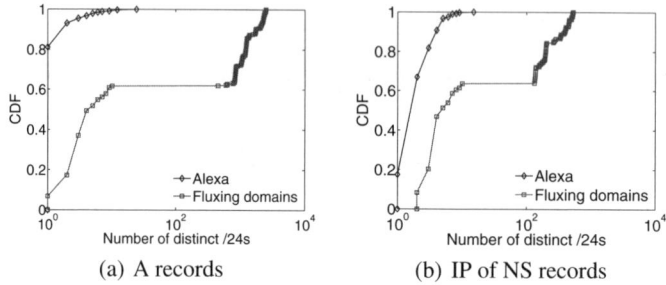

Fig. 6. Distribution of unique /24s that appeared for *all* records in a reply

previous studies [13, 5], we observed many different types of changes in addition to single flux (A records) and double flux (A, and IP address of NS). Another notable point is that each campaign tends to combine techniques: For Pharmacy-A, 52% of domains are double flux and 48% change all three types of records. This result indicates that a single campaign can operate using multiple infrastructures.

4   Location

In this section, we examine the network and geographic locations of hosts that are hosting scam Web sites or serving as name servers; we also compare these locations to those of both spamming hosts and legitimate Web sites hosts.

Topological Location. To examine whether scam sites use different portions of the IP space than the top 500 domains, we studied the distribution of the IPs across the whole IP range. Figure 5 shows that scam networks use a different portion of the IP space than sites that host popular legitimate content. The IPs that host legitimate sites are considerably more distributed. More than 30% of these sites are hosted in the 30/8-60/8 IP address range, which hosted almost none of the scam sites observed in our study:

- *The predominant networks that host scam sites differ from those that host spammers for the corresponding scam campaigns (Figure 5).* Our technical report lists for the top ten ASes by the number of IP addresses for A records (i.e., hosting sites), NS records (i.e., nameservers), and spammers (as observed in the spam trap) [7]. Interestingly, there is almost no overlap between the ASes hosting the scam sites and the nameservers (mostly Asia) and the ASes hosting the spamming IP addresses (mostly Latin America, Turkey, and US). The fact that significant differences exist between networks of scam infrastructure and those of spammers suggest that hosts in different regions of the IP address space do in fact play different "roles" in spam campaigns.

– *DNS lookups for scam domains often return much more widely distributed IP addresses than lookups for legitimate Web sites.* Our intuition was that fast-flux networks that hosted scam sites would be more distributed across the network than legitimate Web hosting sites, particularly from the perspective of DNS queries from a single client (even in the case of a distributed content distribution network, DNS queries typically map a single client to a nearby Web cache). Figure 6 shows the distribution of distinct /24s that appear at the answer section of the DNS replies) for all records in the reply. Roughly 40% of all A records returned for scam domains were distributed across at least 300 distinct /24s, and many were distributed across thousands of /24s. An overly widespread distribution of query replies may indicate that a domain is indeed suspicious (e.g., a fast-flux network).

Geographic Location. Hosting servers and name servers are widely distributed. In total, we observed IP addresses for A records in 283 ASes across 50 countries, IP addresses for NS records in 191 ASes across 40 countries, and IP addresses for spammers across 2,976 IP addresses across 157 countries. Although many scam nodes appear to be in Russia, Germany, and the US, the long list of ASes and countries shows that scam networks are truly distributed; this geographical distribution may be necessary to accommodate the diurnal pattern of compromised hosts' uptime [3]. Interestingly, the countries that are referred to by the most A records are not the same set of countries that host authoritative nameservers for those domains (as indicated by IP addresses of NS records). In particular, Slovakia, Israel, and Romania appear to host more nameservers than sites, and China appears to host relatively more nameservers. This difference in distribution deserves further study; one possible explanation is that nameserver infrastructure for fast-flux networks must be more robust than the sites that host scams (which might be relatively transient).

5   Summary

This paper studied dynamics and roles of fast-flux networks in mounting scam campaigns. We actively monitored the DNS records for URLs for scam campaigns received at a large spam sinkhole over a one-month period to study dynamics features of fast-flux service networks as they are used to host online scam and contrast to the dynamics used for load balancing for popular Web sites. Our findings suggest that monitoring the infrastructure for unusual, invariant changes in DNS mappings may be helpful for automating detection. We plan to explore this possibility in future work.

References

1. Alexa. Alexa the Web Information Company (2008), http://www.alexa.com/
2. Anderson, D.S., Fleizach, C., Savage, S., Voelker, G.M.: Spamscatter: Characterizing Internet Scam Hosting Infrastructure. In: USENIX Security Symposium (August 2007)

3. Dagon, D., Zou, C., Lee, W.: Modeling Botnet Propagation Using Time Zones. In: The 13th Annual Network and Distributed System Security Symposium (NDSS 2006), San Diego, CA (February 2006)
4. Holz, T., Corecki, C., Rieck, K., Freiling, F.C.: Measuring and Detecting Fast-Flux Service Networks. In: NDSS (February 2008)
5. ICANN Security and Stability Advisory Committee. SSAC Advisory on Fast Flux Hosting and DNS (March 2008), http://www.icann.org/committees/security/sac025.pdf
6. Jung, J., Sit, E.: An Empirical Study of Spam Traffic and the Use of DNS Black Lists. In: Internet Measurement Conference, Taormina, Italy (October 2004)
7. Konte, M., Feamster, N., Jung, J.: Fast Flux Service Networks: Dynamics and Roles in Online Scam Hosting Infrastructure. Technical Report GT-CS-08-07 (September 2008), http://www.cc.gatech.edu/~feamster/papers/fastflux-tr08.pdf
8. Passerini, E., Paleari, R., Martignoni, L., Bruschi, D.: FluXOR: detecting and monitoring fast-flux service networks. In: Zamboni, D. (ed.) DIMVA 2008. LNCS, vol. 5137, pp. 186–206. Springer, Heidelberg (2008)
9. Pathak, A., Hu, Y.C., Mao, Z.M.: Peeking into Spammer Behavior from a Unique Vantage Point. In: First USENIX Workshop on Large-Scale Exploits and Emergent Threats (LEET), San Francisco, CA (April 2008)
10. Rajab, M., Zarfoss, J., Monrose, F., Terzis, A.: A Multifaceted Approach to Understanding the Botnet Phenomenon. In: ACM SIGCOMM/USENIX Internet Measurement Conference, Brazil (October 2006)
11. Ramachandran, A., Feamster, N.: Understanding the Network-Level Behavior of Spammers. In: SIGCOMM (September 2006)
12. Spam Trackers, http://spamtrackers.eu/wiki/index.php?title=Main_Page
13. The Honeynet Project. Know Your Enemy: Fast-Flux Service Networks (July 2007), http://www.honeynet.org/papers/ff/
14. Xie, Y., Yu, F., Achan, K., Gillum, E., Goldszmidt, M., Wobber, T.: How dynamic are IP addresses? In: ACM SIGCOMM, Kyoto, Japan (August 2007)
15. Zdrnja, B., Brownlee, N., Wessels, D.: Passive monitoring of DNS anomalies. In: Hämmerli, B.M., Sommer, R. (eds.) DIMVA 2007. LNCS, vol. 4579, pp. 129–139. Springer, Heidelberg (2007)

Inferring Spammers in the Network Core

Dominik Schatzmann, Martin Burkhart, and Thrasyvoulos Spyropoulos

Computer Engineering and Networks Laboratory, ETH Zurich, Switzerland
{schatzmann,burkhart,spyropoulos}@tik.ee.ethz.ch

Abstract. Despite a large amount of effort devoted in the past years trying to limit unsolicited mail, spam is still a major global concern. Content-analysis techniques and blacklists, the most popular methods used to identify and block spam, are beginning to lose their edge in the battle. We argue here that one not only needs to look into the network-related characteristics of spam traffic, as has been recently suggested, but also to look deeper into the network core, to counter the increasing sophistication of spammers. At the same time, local knowledge available at a given server can often be irreplaceable in identifying specific spammers.

To this end, in this paper we show how the local intelligence of mail servers can be gathered and correlated *passively*, scalably, and with low-processing cost at the ISP-level providing valuable network-wide information. First, we use a large network flow trace from a major national ISP, to demonstrate that the pre-filtering decisions and thus spammer-related knowledge of individual mail servers can be easily and accurately tracked and combined at the flow level. Then, we argue that such aggregated knowledge not only allows ISPs to monitor remotely what their "own" servers are doing, but also to develop new methods for fighting spam.

1 Introduction

According to IronPort's 2008 Security Trend Report [1], as much as 90% of inbound mail is spam today. Moreover, spam is no longer simply an irritant but becomes increasingly dangerous. 83% of spam contains a URL. Thus, phishing sites and trojan infections of office and home systems alike are just one click away. The rapid increase of spam traffic over the last years poses significant processing, storage, and scalability challenges for end-host systems, creating a need to at least perform some fast "pre-filtering" on the email server level. To do this, email servers evaluate information received at various steps of the SMTP session using local (e.g., user database, greylisting [2]) and global knowledge (e.g., blacklists [3,4] or SPF [5]) to identify and reject malicious messages, without the need to look at the content.

Nevertheless, traditional pre-filtering methods like blacklists are starting to lose their edge in the battle. Spammers can easily manipulate an IP block for a short time to do enough damage before they can be reported in a blacklist [6,7]. To amend this, new filtering approaches focusing on general network-level characteristics of spammers are developed [8,9,10,11], which are more difficult for a spammer to manipulate. An example of such characteristics are geodesic distance between sender and recipient [12], round trip time [9] or MTA link graph properties [13,14]. These methods have been shown to successfully unveil additional malicious traffic that slips under the radar of

S.B. Moon et al. (Eds.): PAM 2009, LNCS 5448, pp. 229–238, 2009.
© Springer-Verlag Berlin Heidelberg 2009

traditional pre-filtering. Yet, they require different amounts of information and processing, ranging from simply peeking into a few entries of the packet header to less lightweight, more intrusive approaches.

Our work is in the same spirit, in that we are also interested in the network-level characteristics of spammers. However, we look at the problem from a somewhat different perspective. Specifically, we look at the problem from an AS or ISP point of view comprising a network with a large number of email servers. We assume that a number of servers in this network (if not all) already perform *some* level of pre-filtering, e.g. dropping a session to an unknown recipient, using a blacklist, or even using sophisticated network characteristics based mechanisms like the one proposed in [12]. This essentially implies that (a) each server is not performing equally "well" in identifying and blocking spammers, and (b) each server has a limited, *local* view or opinion about which senders are suspicious or malicious. In this context, we're interested in answering the following question: *can one use a 100% passive, minimally intrusive, and scalable network-level method to (a) infer and monitor the pre-filtering performance and/or policy of individual servers, and (b) collect and combine local server knowledge in order to re-use it to improve server performance?*

Although one could potentially use individual server logs to gain the needed pre-filtering information, in order to collect network-wide spam statistics an ISP would have to gather the logs of all mail servers in the network. As these servers are usually located in many different organizational domains, this is a tedious process that is hindered by privacy concerns of server operators. Instead, *we demonstrate that the pre-filtering decisions of individual servers can be passively and accurately inferred in the network using little flow size information captured in the network core* as illustrated in Fig. 1. Having validated this methodology, we then use it to analyze the incoming SMTP traffic of a major national ISP network with 320 internal email servers. We found that internal servers perform very differently. Some servers accept up to 90% of all incoming SMTP flows, while many accept only $10 - 20\%$. We look further into the causes of these discrepancies, and after ruling out various "benign" causes, we conclude that many servers in the network seem to be mis-configured or simply under-performing. Based on this, we investigate how and to what extent the *collective* knowledge of well-performing servers could be used to improve the pre-filtering performance of everyone.

Summarizing, our method avoids the cumbersome process of log gathering and correlation. It also requires minimal processing and session information, implying that this method is scalable enough to keep up with the high amount of information constantly gathered at the network core. Finally, it is complementary to recently proposed, sophisticated spam detection mechanisms based on network characteristics, in that the whole system could benefit from such increased capabilities deployed a given server or subset of them.

2   Preliminaries

The email reception process on a server consists of three phases as depicted in Fig. 2, TCP handshake, SMTP email envelope exchange, and email data exchange. Pre-filtering is employed in the second phase: in order to identify and quickly reject malicious traffic

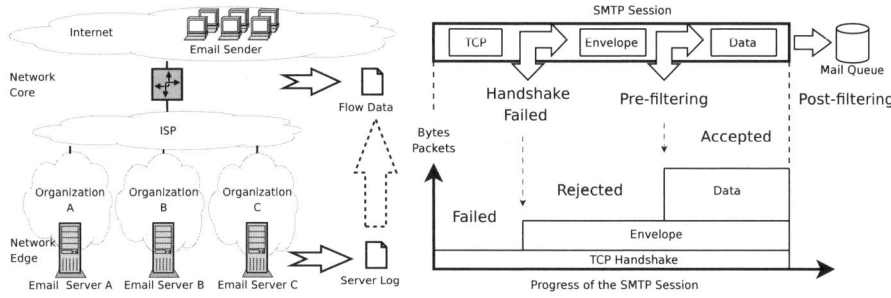

Fig. 1. The ISP view of the network   **Fig. 2.** The three phases of email reception

based on the message envelope only a server may use "global" knowledge (e.g., sender listed in a blacklist), local knowledge (e.g., attempt to reach unknown recipients), or policy-based decisions (e.g., greylisting).

We analyzed the log of a university mail server serving around 2400 user accounts and receiving on average 2800 SMTP flows per hour to look into such pre-filtering performance in more detail. We found that as much as 78.3% of the sessions were rejected in the pre-filtering phase. 45% of the rejects were based on local information (e.g., user database or greylisting) and only 37.5% were due to blacklists. This illustrates the importance of local mail server knowledge for spam detection.

Based on the server's decisions, we classify observed SMTP sessions as either *failed*, *rejected* or *accepted*. Our key observation is that, whenever a sender manages to get to the next phase, the overall transferred information is significantly increased. For example, if a sender is accepted and allowed to send email content, he is able to transmit much more data than a sender already rejected in phase two. As a consequence, we conjecture that flow properties reflecting the size or length of SMTP sessions, such as the flow size or packet count, should be an accurate indicator for the phase in which an SMTP session was closed.

We validate this assumption in Section 3. For this purpose, we have used three weeks of unsampled NetFlow data from January, February and September 2008 (referred to as week 1, 2, 3), captured at the border routers of a major national ISP [15] serving more than 30 universities and government institutions. The IP address range contains about 2.2 million internal IP addresses and the traffic volume varies between 60 and 140 million NetFlow records per hour. The identification of SMTP traffic is based on TCP destination port 25[1]. Based on the SMTP traffic, a list of active internal email servers was generated and verified by active probing. We detected 320 internal servers, receiving up to 2 million SMTP flows per hour.

3   SMTP Flow Characteristics

In this Section, we demonstrate how the effect of pre-filtering on flow characteristics can be used to track the servers' decisions for each SMTP session.

---

[1] Note that only the traffic flowing from external SMTP clients to internal servers is considered.

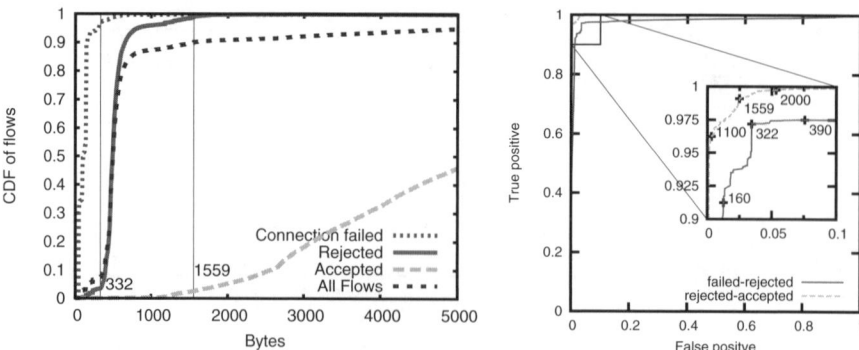

Fig. 3. Byte count distribution   **Fig. 4.** ROC curve for bytes per flow metric

Table 1. Classification performance for x bytes per flow

	$x < 322$	$322 <= x <= 1559$	$x > 1559$
Failed	9302 (95.64%)	417 (4.28%)	7 (0.07%)
Rejected	11008 (3.59%)	409675 (96.66%)	3132 (0.74%)
Accepted	55 (0.09%)	1662 (2.74%)	58845 (97.16%)

The CDF of byte counts for flows arriving at the mail server in week 1 is presented in Fig 3. The class of failed connections mainly consists of very small flows as only a small number of packets could be sent. 97% of these flows have less than 322 bytes. The size of most rejected flows is between 400 and 800 bytes. This corresponds to the size of the SMTP envelope. Lastly, the distribution of accepted flow sizes is dominated by the overall email size distribution and reaches 50% at around 5000 bytes. This is consistent with the findings of Gomes et al. [16]. The CDF for "all flows" in Fig. 3 is a superposition of the three classes weighted by their relative probability of appearance. All three classes are well visible in the total CDF even though it is dominated by rejected flows due to the fact that around 80% of all flows are rejected.

Next, we determined two optimal threshold sizes to differentiate between *rejected*, *failed* and *accepted* flows. For this purpose, we constructed ROC curves [17] which plot the true positive versus the false positive rate of a detector for a range of thresholds. Fig. 4 shows the ROC curves for the detection of rejected vs. failed and accepted vs. rejected flows. The three classes are distinguishable with high precision. We selected the two thresholds 332 Bytes (rejected vs. failed) and 1559 Bytes because these points are closest to the top left corner and hence yield the best detection quality [17].

We evaluated the false positive rate of these threshold detectors on data of another week (week 2) and present the results in Table 1. The false detection rate is below 4.5% for all classes which is sufficiently accurate for the applications outlined in Section 4[2]. We also analyzed the power of other flow properties to discriminate between the three classes. In addition to *bytes per flow*, also *packets per flow* and *average bytes per packet* are well suited for this purpose [18].

---

[2] The flow labels assigned by our system are to be treated mostly as "soft" labels. Further accuracy could be achieved by using e.g. clustering algorithms on additional flow fields.

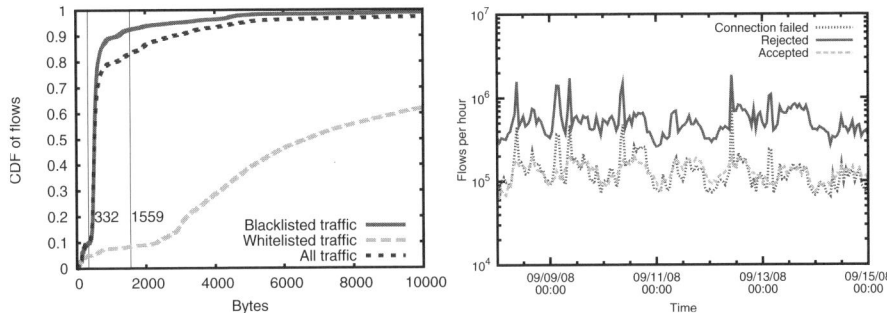

Fig. 5. Network-wide flow sizes   **Fig. 6.** Network-wide pre-filtering statistics

It is important to note that packet sampling would affect our approach. Over 90% of the rejected SMTP sessions consist of 10 or less packets and more than 90% of accepted sessions have less than 40 packets. With a sampling rate of 1:100 or even 1:1000, the resulting flows would mostly consist of 1 or 2 packets. This would weaken the usefulness of the bytes and packets per flow metrics; yet, our analysis suggests that it could still be possible to distinguish between rejected and accepted flows using *average bytes per packet* [18]. Further, adaptive sampling techniques are being developed [19] that could perhaps address this problem also. We intend to look further into the issue of sampling in future work.

Network-wide characteristics. The classification of flows based on their size allows to passively monitor pre-filtering activity in large networks in a scalable manner, without resorting to server logs. To validate that the observed characteristics also hold on a network-wide scale, we show the characteristics of black- and whitelisted traffic for the 50 most active mail servers in our network in Fig. 5. The shape of black-/whitelisted curves nicely reflects the characteristics of rejected and accepted flows from Fig. 3. Hence, (i) the vast majority of traffic from blacklisted hosts is rejected by our network in pre-filtering and (ii) we are able to infer this reject decisions from flows sizes only. Individual server performance differences are addressed in detail in Section 4.1.

The generation of network-wide pre-filtering statistics, as illustrated in Fig. 6, allows to easily estimate the amount and track the dynamics of incoming spam at the ISP level. An ISP is now able to investigate the root cause of anomalies in rejected/accepted traffic. Potential causes are global spam campaigns that are visible on many servers, spamming attacks targeted to a small set of servers or misconfiguration and performance problems of single servers.

4   Applications

We now turn our attention to potential applications of our method. In Section 4.1, we demonstrate how it can be used to passively analyze the configuration of mail servers and troubleshoot misconfigured servers. We then explore the feasibility and potential of a collaborative filtering system among the various mail servers in Section 4.2.

4.1 Email Server Behavior

Today, adequate configuration and maintenance of mail servers is a time-consuming process. It would be very helpful for operators to get a performance map of the various mail servers present in the network. The state of pre-filtering deployment in the network could be checked regularly and potential configuration problems, (e.g., the presence of open relay servers), could be addressed proactively.

To compare the pre-filtering performance of internal servers, we define the *acceptance ratio* of a server to be the number of accepted SMTP flows divided by the number of total SMTP flows seen by the server. A high ratio of, for example, 0.9 indicates that 90% of all incoming SMTP sessions are accepted, whereas a low ratio indicates that most of the connections are rejected during the TCP handshake or the SMTP envelope. Clearly, the observed acceptance ratio for a server is affected by two parameters: (i) the *traffic mix* of ham and spam for this server, and (ii) the server *prefiltering policy*. To address the former, we estimated the spam/ham mix ratio for each server with the help of the XBL blacklist from Spamhaus. Our analysis shows that spam (flows from blacklisted sources) is evenly distributed among servers. 81% of the servers have a spam load between 70% and 90%, consistent with [1]. This results implies that *big differences in servers' acceptance ratios cannot be attributed to different traffic mixes.*

The server policy issue is somewhat trickier. The above numbers imply that, if all servers were at least using a blacklist, the acceptance ratio of most internal servers should be between 0.1 and 0.3, with differences attributed to traffic mix and sophistication and/or aggressiveness of pre-filtering policies (e.g., greylisting, etc.). Instead, the acceptance ratios of the top 200 servers for week 3 of our data set range from 0.003 up to 0.93 with a mean of 0.33 as can bee seen in Fig. 7. 35% of the servers have an acceptance ratio > 0.30. Based on the above traffic mix estimation, we conclude that they are accepting a lot of traffic from spam sources. This could imply: (i) a regular server that is sub-optimally configured, lacks sophisticated or even simple pre-filtering measures (e.g., lack of time, caring, or knowhow), and/or should at least raise an eyebrow; or (ii) a server whose intended policy is to accept all messages (e.g., servers that apply content-based filtering only, honeypots, etc.)

To verify this assumption, we sent emails to all servers from two different IP addresses: an address blacklisted by Spamhaus and Spamcop and an ordinary, not blacklisted address[3]. The reaction of the servers to our sending attempts clarified whether the server was using greylisting and/or blacklisting. The servers classified as 'unknown' are those servers for which the reaction was not conclusive. The high concentration of black- and greylisting servers below the average ratio shows that, indeed, these servers implement basic pre-filtering techniques, whereas servers that do not implement them lie mostly above average. Also, with increasing volume (to the right), servers with high acceptance ratios tend to disappear. This affirms that administrators of high-volume servers (have to) rely on aggressive pre-filtering to master the flood of incoming mails. We also manually checked high acceptance servers and found no honeypots trying to deliberately attract and collect spam.

---

[3] It is important to stress that this, and other "manual" investigations we performed in this section are only done for validation purposes, and are not part of the proposed system.

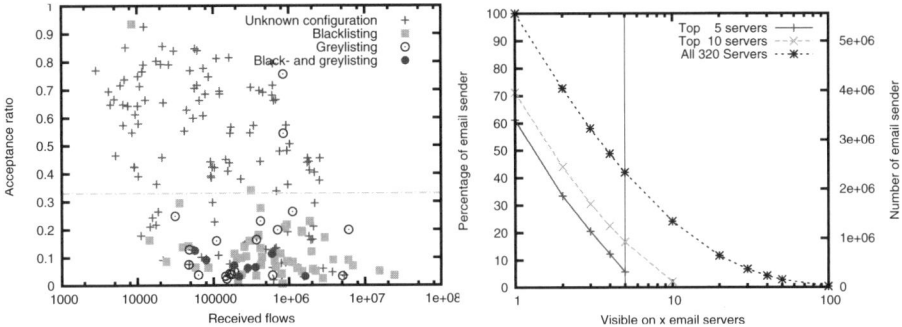

Fig. 7. Server acceptance ratios vs. traffic volume   **Fig. 8.** Visibility of the email senders

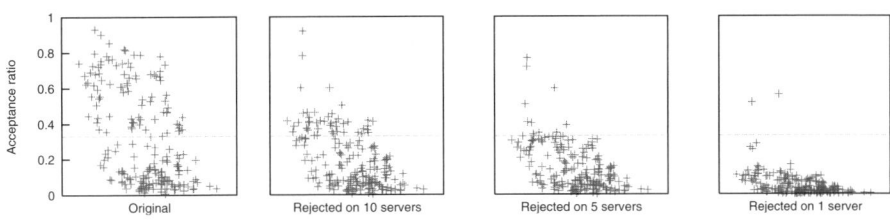

Fig. 9. Improvement potential when using collaborative filtering

We conclude that differences in acceptance ratio are mainly due to configuration issues and that there is a large group of servers that might need a wake-up call or could profit from the expertise of other servers.

4.2 Collaborative Filtering

Given the above observations of unbalanced mail server performance, what could an ISP do to improve overall pre-filtering performance in its network? Surely, the ISP could contact individual administrators and organize a meeting to present the statistics about mail server performance where administrators would exchange their knowhow. However, this requires a lot of organizational effort, needs to take place regularly, and attendance of administrators is not guaranteed. Furthermore, educating customers' email administrators is usually not the ISP's business.

Therefore we investigate a passive way enabling *all* servers to profit from the local knowledge and the best existing anti-spam techniques already present in *some* servers in the network. By accepting more or less emails from a client, email servers actually perform an implicit rating of the very client. With our method, these ratings could be extracted from traffic and used to build a collaborative filtering system (CFS), or more properly, a collaborative rating/reputation system. The system would recommend accepting/rejecting mails from a client, based on the behavior of the collective of all servers: "This host was rejected by 80% of the servers in the network. Probably you should reject it as well."

It is important to note that the added value of the collected information depends on the ability of different servers to block different flows: heterogeneous knowledge or heterogeneous pre-filtering detection techniques and policies are *desirable* since this increases the chance that at least some methods will detect the spammer.

At the same time, a significant overlap of the sets of email senders visible to each server is needed, in order to achieve useful and trusted enough ratings. We analyzed the visibility of the sending hosts on three sets of top 5, top 10 and all internal servers (see Fig. 8). More than 5.5 million email senders are visible on the 320 internal servers. Moreover, 72% of all email senders are visible on at least two internal servers. This means that for 72% of the email senders a second option from another server is available. Note that even the top 5 email servers only see 61% of the sending hosts. In addition, only 6% of the email senders are visible on all of these five servers. This percentage is increased to 17% for the top 10 servers. By using information from all internal servers, 42% of the sending hosts are visible on at least 5 different servers, which is a promising foundation for building a CFS. In addition we explicitly analyzed the visibility of hosts that are listed in a black- or whitelist. The visibility of blacklisted hosts follows the overall host visibility. However, the visibility of whitelisted hosts is even better (e.g., 60% of the whitelisted hosts are visible on at least 5 different servers).

The actual implementation of a CFS is beyond the scope of this paper. Nevertheless, we are interested here in estimating the potential of such a system by simulating simple filter rules. Specifically, we simulated an *online* version of the system where blocklists are created as incoming flows are parsed according to some simple rules. Specifically, counters are maintained for senders that have been consistently blocked with the following 3 rules:[4] As soon as a sender's connections have been blocked by at least 10, 5 or 1 server(s), respectively, the sender is entered into our blocklist. Then, if a sender is on the list all incoming connections by this sender will be counted as rejected. The resulting acceptance ratios assuming all servers are using the CFS are shown in Fig 9.

There is an inherent tradeoff in the number of server "votes" used to make a decision. By requiring many rejecting servers (e.g., 10) for membership in the blocklist, the reliability of the filtering becomes quite strong (only 3 blocked hosts were actually whitelisted). Yet, the size of the blocklist is reduced, and provides less additional information than blacklists. Specifically, in the 10 server case, the blocklist has 83% overlap with the Spamhaus blacklist and could be used to block 47% of all SMTP sessions. In the other extreme, if only one rejecting server is required to be put on the blocklist, acceptance ratios of all servers are dramatically reduced. Further, the blocklist overlap with Spamhaus is only 70% and could be used to block up to 95% of all SMTP connections. However, 185 members were whitelisted. That is, requiring only one rejecting server introduces a higher false rejection rate. Nevertheless, it is important to note that in both cases, a significant amount of hosts identified by this method are *not* found in the blacklist, underlining the collaborative nature of the system, and implying that simply making under-performing servers use a blacklist, would not suffice.

Concluding, our estimation shows that there is a significant information overlap that can be leveraged to improve overall pre-filtering by making local server information accessible to the entire network, in an almost seamless manner. Today, pre-filtering is dominated by the use of DNSL blacklists. However, the CFS is independent of the very techniques applied and will automatically profit from upcoming improved techniques.

---

[4] To reduce the effects of greylisting we delayed the blacklisting process by 15 min.

As a final note, due to the inherent "softness" of the labels attached by such a flow-based system, and the various risks of blocking a session in the network core, we stress here that our system is not intended as an actual blocking filter, but rather as a reputation rating system, which individual email servers can *opt* to include in their pre-filtering phase. Consequently, servers whose policy is to accept all emails, do not have to be affected by our system, although using it, could perhaps provide hints to the content-based filter.

5   Discussion

Although being only a first step, we believe the proposed method has important potential to be applied in production networks and also paves the way for future research in the area of network-level and network-wide characteristics of spam traffic. In this Section we discuss some possible limitations of our approach.

Delay: The flow data is exported by the router only after the TCP connection is completed. Therefore, the information about the flow is delayed at least until the session is closed. We measured this delay to be less than 7.7 seconds for 90% of all SMTP flows. This illustrates that any application based on flow characteristics is limited to near-realtime. In particular, properties of a flow can not be used to intercept this very flow. For our proposed applications, this limitation is acceptable as we are not interested in using the properties of a flow to intercept this very flow, but rather subsequent ones.

Flow size manipulation: In principle, spammers could adapt to our method by prolonging the envelope phase, for instance by sending multiple RCPT or HELO commands. This would indeed increase the number of transmitted bytes per flow but will, at the same time, increase the number of packets. However, the average number of bytes per packet remains small and the bytes per packet metric could still be used for the classification. Further, the spammer could try to increase the bytes transmitted in each command by using long email addresses, but maximum size for a command is limited to 521 characters [20]. Moreover, any deviation from normal SMTP behavior could easily be detected and mail servers could enforce the shortness of the envelope phase as there is no need to be overly long in a normal use case. In addition, the misbehavior of the host could be published to make this information available for other servers.

In a more sophisticated scenario, a spammer could take over different internal hosts and reconfigure them as internal mail servers. By sending accepted SMTP traffic from bot to bot, one could try to positively influence the CFS ratings for these bots. The design of the CFS needs to be aware of this problem. In a first step, only servers that have been active over a longer time period (i.e., they have been accepting mails for at least several weeks) and get a certain amount of connections from trusted email servers (e.g., on a whitelist) could be included into the filtering process.

6   Conclusion

Mail server administrators are engaged in an arms race against spammers. They urgently need new approaches to fight state-of-the-art attachment spam increasingly originating from low-profile botnet spammers. In this paper, we demonstrated that simple flow metrics, such as byte count, packet count, and bytes per packet, successfully discriminate between spam and ham flows when pre-filtering is deployed in mail servers. Thus, one could infer individual mail server's decisions with respect to the legitimacy and acceptance of a given connection. This allows an operator i) to concentrate dispersed mail server knowledge at the network core and ii) to passively accumulate network-wide spam statistics, profile filtering performance of servers, and rate clients. Thus, the advantages of flow and server log analysis finally meet at the network core. We believe this is an important step towards successfully fighting spammers at the network-level.

References

1. IRONPORT: 2008 internet security trends, http://www.ironport.com
2. Harris, E.: The next step in the spam control war: Greylisting (2003)
3. SpamCop: Spamcop blocking list, http://www.spamcop.net/bl.shtml
4. Spamhaus: The spamhaus block list, http://www.spamhaus.org/sbl
5. Wong, M., Schlitt, W.: Sender Policy Framework (SPF). RFC 4408
6. Ramachandran, A., Dagon, D., Feamster, N.: Can DNS-based blacklists keep up with bots. In: Conference on Email and Anti-Spam, CEAS 2006 (2006)
7. Duan, Z., Gopalan, K., Yuan, X.: Behavioral Characteristics of Spammers and Their Network Reachability Properties. In: IEEE International Conference on Communications, ICC 2007 (2007)
8. Ramachandran, A., Feamster, N., Vempala, S.: Filtering Spam with Behavioral Blacklisting. In: ACM conference on Computer and Communications Security, CCS 2007 (2007)
9. Beverly, R., Sollins, K.: Exploiting Transport-Level Characteristics of Spam. In: CEAS 2008 (2008)
10. Clayton, R.: Using Early Results from the spamHINTS. In: CEAS 2006 (2006)
11. Gu, G., Perdisci, R., Zhang, J., Lee, W.: BotMiner: Clustering Analysis of Network Traffic for Protocol-and Structure-Independent Botnet Detection. In: USENIX Security Symposium (July 2008)
12. Syed, N.A., Feamster, N., Gray, A., Krasser, S.: Snare: Spatio-temporal network-level automatic reputation engine. Technical Report GT-CSE-08-02, Georgia Tech. (2008)
13. Desikan, P., Srivastava, J.: Analyzing network traffic to detect e-mail spamming machines. In: ICDM Workshop on Privacy and Security Aspects of Data Mining (2004)
14. Gomes, L.H., Almeida, R.B., Bettencourt, L.M.A., Almeida, V., Almeida, J.M.: Comparative Graph Theoretical Characterization of Networks of Spam and Legitimate Email. Arxiv physics/0504025 (2005)
15. SWITCH: The swiss education and research network, http://www.switch.ch
16. Gomes, L.H., Cazita, C., Almeida, J.M., Almeida, V., Meira, W.: Characterizing a spam traffic. In: ACM SIGCOMM conference on Internet measurement, IMC 2004 (2004)
17. Fawcett, T.: An introduction to roc analysis. Pattern Recognition Letters 27 (2006)
18. Schatzmann, D., Burkhart, M., Spyropoulos, T.: Flow-level characteristics of spam and ham. Technical Report 291, Computer Engineering and Networks Laboratory, ETH Zurich (2008)
19. Ramachandran, A., Seetharaman, S., Feamster, N., Vazirani, V.: Fast monitoring of traffic subpopulations. In: ACM SIGCOMM Conference on Internet Measurement, IMC 2008 (2008)
20. Klensin, J.: Simple mail transfer protocol. RFC 2821 (April 2001)

Beyond Shannon: Characterizing Internet Traffic with Generalized Entropy Metrics

Bernhard Tellenbach[1], Martin Burkhart[1], Didier Sornette[2], and Thomas Maillart[2]

[1] Computer Engineering and Networks Laboratory, ETH Zurich, Switzerland
[2] Department of Management, Technology and Economics, ETH Zurich, Switzerland
{betellen,martibur,dsornette,tmaillart}@ethz.ch

Abstract. Tracking changes in feature distributions is very important in the domain of network anomaly detection. Unfortunately, these distributions consist of thousands or even millions of data points. This makes tracking, storing and visualizing changes over time a difficult task. A standard technique for capturing and describing distributions in a compact form is the Shannon entropy analysis. Its use for detecting network anomalies has been studied in-depth and several anomaly detection approaches have applied it with considerable success. However, reducing the information about a distribution to a single number deletes important information such as the nature of the change or it might lead to overlooking a large amount of anomalies entirely. In this paper, we show that a generalized form of entropy is better suited to capture changes in traffic features, by exploring different moments. We introduce the Traffic Entropy Spectrum (TES) to analyze changes in traffic feature distributions and demonstrate its ability to characterize the structure of anomalies using traffic traces from a large ISP.

1 Introduction

Fast and accurate detection of network traffic anomalies is a key factor in providing a reliable and stable network infrastructure. In recent years, a wide variety of advanced methods and tools have been developed to improve existing alerting and visualization systems. Some of these methods and tools focus on analyzing anomalies based on volume metrics, such as e.g., traffic volume, connection count or packet count [1]; others look at changes in traffic feature distributions [2] or apply methods involving the analysis of content or the behavior of each host or group of hosts [3]. However, content inspection or storing state information on a per host basis are usually limited to small- and medium-scale networks. If feasible at all, the link speeds and traffic volumes in large-scale networks hinder a reasonable return on investment from such methods. Most approaches designed for large-scale networks have therefore two things in common: First, they reduce the amount of input data by looking at flow-level information only (e.g., Cisco NetFlow [4] or IPFIX [5]). Second, they use on-the-fly methods that do not rely on a large amount of stored state information. A major drawback of on-the-fly methods is their inappropriateness for approaches relying on the history of traffic feature distributions. A related problem arises, when one wants to visualize the evolution of IP address- or flow size distributions over time. In large-scale networks, these distributions consist of millions of data points and it is unclear how to select a relevant subset.

S.B. Moon et al. (Eds.): PAM 2009, LNCS 5448, pp. 239–248, 2009.

A prominent way of capturing important characteristics of distributions in a compact form is the use of entropy analysis. Entropy analysis (1) reduces the amount of information needed to be kept for detecting distributional changes and (2) allows for a compact visualization of such changes. Evidence that methods based on Shannon entropy capture the relevant changes has been documented [6,7,8].

Here, we propose a new method for capturing and visualizing important characteristics of network activity based on generalized entropy metrics. Our method is a significant extension of the work of Ziviani et al. [9] who recently proposed and studied the use of generalized entropy metrics in the context of anomaly detection. Ziviani et al. introduced a method based on a single generalized entropy value which needs to be tuned to a specific attack. In their evaluation, they provide evidence that generalized entropy metrics are better suited to capture the network traffic characteristics of DoS attacks than Shannon entropy.

Our new method makes the following contributions:

- We define the *Traffic Entropy Spectrum (TES)* for capturing and visualizing important characteristics requiring little or no tuning to specific attacks.
- We demonstrate that the TES can not only be used for the detection of an anomaly but also for capturing and visualizing its characteristics.
- We provide evidence that Autonomous System (AS) entropy is a valuable complement to IP address entropy.
- We confirm the finding of [9] for a broader set of anomalies.

The remainder of this paper is organized as follows: In Section 2, we start with a review of the Tsallis entropy and discuss its advantage over Shannon entropy. Next, we introduce the Traffic Entropy Spectrum (TES) and explain how it is used to capture and visualize distributional changes. Section 3 describes the methodology used for the evaluation. Section 4 discusses the results and outlines how TES could be used to build a detector with integrated anomaly classification. Finally, Section 5 discusses related work and section 6 summarizes the results.

2   The Tsallis Entropy

The Shannon entropy $S_s(X) = -\sum_{i=1}^{n} p_i \cdot \log_2(p_i)$ [10] can be seen as a *logarithm moment* as it is just the expectation of the logarithm of the measure (with a minus sign to get a positive quantity). Given that different *moments* reveal different clues on the distribution, it is clear that using other generalized entropies may reveal different aspects of the data. Two of such generalized entropies relying on *moments* different from the *log-moment* are the Renyi and Tsallis entropies, the latter being an expansion of the former. Here, we use the Tsallis entropy

$$S_q(X) = \frac{1}{q-1}\left(1 - \sum_{i=1}^{n} (p_i)^q\right) \tag{1}$$

as it has a direct interpretation in terms of moments of order q of the distribution and has also enjoyed a vigorous study of its properties [11,12,13,14,15] From these properties, it follows that Tsallis entropy is better suited to deal with non-Gaussian measures, which are well-known to characterize Internet traffic [16,17,2], while Shannon's entropy is better adapted to Normal distributions. Note that this and other interesting aspects of the Tsallis entropy are the reason why it has many applications to complex systems[1].

2.1 Meaning of the Parameter q

When using the Tsallis entropy, there is not one Tsallis entropy but as many as there are possible choices for q. Each q reveals different aspects of distributions used to characterize the system under study. Before we take a closer look at the meaning of q, we need to define what kind of distributions we want to use and how we get it. We start with the definition of important terms used in the reminder of the paper:

- *system*: A (set of) network(s) described by an ensemble of network flows
- *feature*: Any flow property that takes on different values and whose characterization using a distribution is potentially useful. Flow properties used in this study are: source and destination IP address, source and destination port and origin and destination Autonomous System (AS).
- *element i*: A specific instance of a feature (e.g., source IP address 10.0.0.1)
- *activity a_i*: The number of occurrences of element i within a time slot of size T. Slot sizes used for this study are: 5, 10 and 15 minutes.
- *feature distribution*: The probability distribution $P[I = i] = p_i = \frac{a_i}{\sum_j a_j}$ of, e.g., the feature *source port*. Note that p_i can also be interpreted as *relative activity* of i. These feature distributions serve as input for the Tsallis entropy calculation.

We now discuss the meaning of different values of q. First, it is essential to stress that both $q = 0$ and $q = 1$ have a special meaning. For $q = 0$, we get $n - 1$, the number of elements in the feature distribution minus one. For $q = 1$, the Tsallis entropy corresponds to the Shannon entropy. This correspondence can be derived by applying l'Hôpital's rule to (1) for $q \longrightarrow 1$. For other q's, we see that (1) puts more emphasis on those elements which show high (low) activity for $q > 1$ ($q < 1$). Hence, by adapting q, we are able to highlight anomalies that

1. increase or decrease the activity of elements with no or low activity for $q < 1$,
2. affect the activity of a large share of elements for q around 1,
3. increase or decrease the activity of a elements with high activity for $q > 1$.

2.2 The Traffic Entropy Spectrum

To leverage the full capabilities of Tsallis entropy, we introduce a new characterization and visualization method called the Traffic Entropy Spectrum (TES). The TES is a three axis plot that plots the entropy value over time (first axis) and for several values of q (second axis). For convenient 2D presentation, the third axis (showing the normalized entropy values) can be mapped to a color range. Hence, the TES illustrates the temporal dynamics of feature distributions in various regions of activity, ranging from very low activity elements for negative qs to high activity elements for $q > 1$.

But what values should be used for the parameter q and do they need to be tuned to the characteristics of the network traffic at a specific sensor? By experimenting with

---

[1] See http://tsallis.cat.cbpf.br/biblio.htm for a complete bibliography

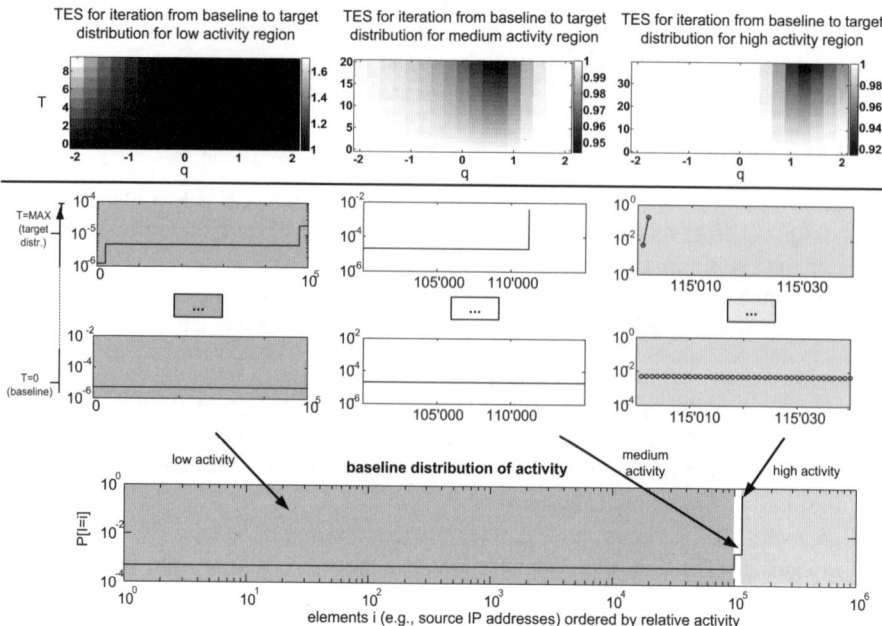

Fig. 1. Impact of changes to different regions of the distribution. Bottom: Baseline and target distributions for low, medium and high activity regions. Top: Resulting TES when altering the distribution in the respective region from the baseline to the target distribution in multiple, even sized, steps.

traces from different sensors and years (2003 to 2008) showing largely differing traffic characteristics, we found that the selection $q = -2, -1.75, ..., 1.75, 2$ gives sufficient information to detect network anomalies in all of those traces. Large values $q > 2$ or smaller values $q < -2$ did not provide notable gains. Hence, this choice of qs worked for many different traces and is therefore strong empirical evidence that it requires little or no tuning to the traffic characteristics of a sensor.

To illustrate the meaning of the parameter q and the TES, we make use of an artificial feature distribution $P[I = i]$ of elements i (see Figure 1) where we identify exactly three different regions. Each region contains elements that show either *low*, *medium*, or *high activity*. Note that for simplicity, all elements in a region have the same absolute activity. We first look at the impact of modifications that are (1) limited to one of those regions and (2) that do not affect the total contribution of this region to $\sum p_i = 1$. To see how the TES reacts to such changes, we specify suitable target distributions for each region (see Figure 1). We then iteratively transform the distribution of a region starting from the baseline distributions in time slot $T = 0$ to the target distributions. We then divide the entropy values we get for the time slots T by the value of the baseline ($T = 0$). Hence, values less than one denote a decrease and values greater than one an increase in entropy compared to the baseline. Figure 1 shows the response of the TES for the transformations of the different regions. Inspecting the TES for the different modifications reveals that they behave as expected:

- high activity: reducing the # of elements decreases entropy for $q > 1$
- medium activity: reducing the # of elements decreases entropy for $-1 < q < 1$
- low activity: reducing the activity of some elements increases entropy for $q < -1$

3 Methodology

For anomaly detection with real traffic traces, we calculated the TES on the activity of the following flow features: Source- and destination IP address, source- and destination ports, origin and destination Autonomous System. We did this for each of the protocols TCP, UDP, ICMP and OTHERS separately.

3.1 Calculating the TES

The calculation of the TES is straightforward. We aggregated the sample distribution of the various traffic features over an interval of 5, 10 and 15 minutes. While the results using the 15 minutes interval are much smoother, shorter intervals are better suited to point out anomalies that last only tens of seconds or a few minutes. At the end of each interval, we calculated the Tsallis entropy values for the different qs and stored them for visualization using the TES. Note that with our selection of qs, we need to store a set of 17 values per interval only.

After calculating the TES, we apply two different normalization methods to compensate for the large absolute difference of the entropies for different q's:

- Global normalization using the maximum and minimum entropy value for a given q during the observation period as follows $S_{normalized,q} = \frac{S_q - minS_q}{maxS_q - minS_q}$. This maps all entropy values to the range [0,1].
- Normalization using the maximum and minimum entropy for a given q on a training day, for instance before the anomaly under scrutiny. Here, we map entropy values between the minimum and maximum of the training day to [0,1]. Other values are either above 1 or below 0.

The TES based on global normalization is used to identify dominating changes. If such a dominating change is present, it stands out at the cost of a decreased visibility of non-dominating changes. The second normalization is used to assess whether changes stay within the variations of the training day. Using the second normalization method, it is easy to develop a simple anomaly detector. Values going below the minimum or above the maximum of the training day, expose the anomalous parts of the TES only. Even though this detection procedure is very straight-forward, our evaluation shows that this simple method is already sufficient for detecting and classifying critical anomalies in network traces.

3.2 Anomaly Characterization

Malicious attacks often exhibit very specific traffic characteristics that induce changes in feature distributions known to be heavy-tailed. In particular, the set of involved values per feature (IP addresses or ports) is often found to be either very small or very large. In a DDoS attack, for instance, the victim is usually a single entity, e.g., a host or a router. The attacking hosts, on the other hand, are large in numbers, especially if source addresses are spoofed. Similarly, if a specific service is targeted by an attack, a single destination port is used, whereas source ports are usually selected randomly. In general, specific selection of victims or services leads to *concentration* on a feature and, in turn, to a change in the high activity domain. In contrast, random feature selection results in *dispersion* and impacts the low activity domain (e.g., spoofed IP addresses only occur once in the trace). Knowing this, it is possible to profile an attack based on the affected activity regions for each feature.

4   Application on Network Traces

The data used in this study was captured from the five border routers of the Swiss Academic and Research Network (SWITCH, AS 559) [18], a medium-sized backbone operator that connects several universities and research labs (e.g., IBM, CERN) to the Internet. The SWITCH IP address range contains about 2.4 million IP addresses, and the traffic volume varies between 60 and 140 million NetFlow records per hour. The records are collected from five different border routers which do not apply any sampling or anonymization. We study the effect of TES using five well-understood events:

- **Refl. DDoS:** A reflector DDoS attack involving 30,000 reflectors within the SWITCH network, used to attack a web server. Two weeks of traffic were analyzed including some preliminary scanning activity (April 2008). Figure 2(a) shows the TES for incoming DstIPs. The attack is clearly visible around 04/11 and lasts for almost one day. Figure 2(b) shows the effective activity of the reflectors during a two-week period. The sustained activity on 04/04 and 04/05 without attack flows suggests that attackers are scanning the network for potential reflectors.
- **DDoS 1:** A short (10 min.) DDoS attack on a router and a host with 8 million spoofed source addresses (Sept. 2007). DstPort is TCP 80. Figure 2(c) plots the TES for incoming Autonomous System (AS) numbers. The attack is nicely visible for $q < 0$ on the 09/01. Although the covered period is 8 days, the attack is visible with an excellent signal to noise ratio and *no false alarms*. Note that for Shannon entropy ($q = 1$) the peak is insignificant.
- **DDoS 2:** A long (13h) DDoS attack on a host with 5 million spoofed source addresses (Dec. 2007/Jan. 2008). DstPort is TCP 80.
- **Blaster Worm:** Massive global worm outbreak caused by random selection/ infection of new hosts, exploiting a RPC vulnerability on TCP DstPort 135 (Aug. 2003).
- **Witty Worm:** Fast spreading worm exploiting a vulnerability in ISS network security products. Uses UDP SrcPort 4000 and random DstPort (March 2004).

4.1   Patterns in Real Traffic

In this Section we analyze the spectrum patterns exhibited by the attacks described previously. For describing these patterns we use a shorthand notation representing the state of S_q with respect to the thresholds by a single character c_q:

$$c_q = \begin{cases} \text{`+'} & \text{if } S_q \geq max\ S_q \text{ of the training day (positive alert)} \\ \text{`-'} & \text{if } S_q \leq min\ S_q \text{ of the training day (negative alert)} \\ \text{`0'} & \text{else (normal conditions)} \end{cases}$$

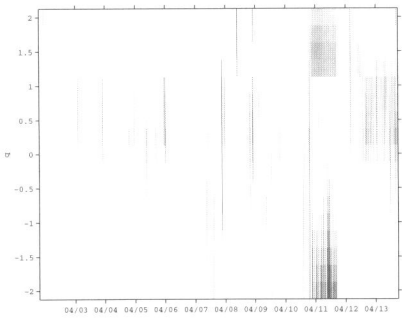

(a) TES of DstIP addresses for flows into our network during the reflector attack. Alerts are shown in red (resp. blue) above (below) threshold of a normal "training day")

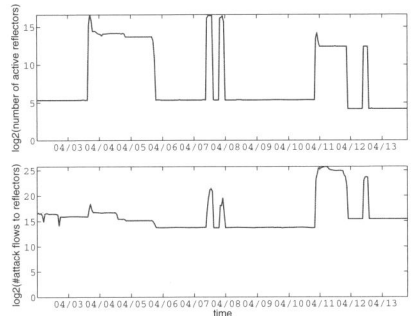

(b) The effective number of active reflectors (top) and the effective number of attack flows toward (candidate) reflectors in our network (bottom)

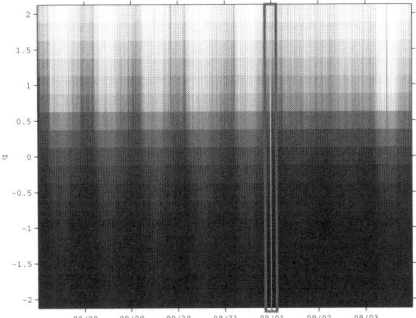

(c) TES of origin autonomous systems in the incoming traffic during the DDoS 1 attack represented with global normalization

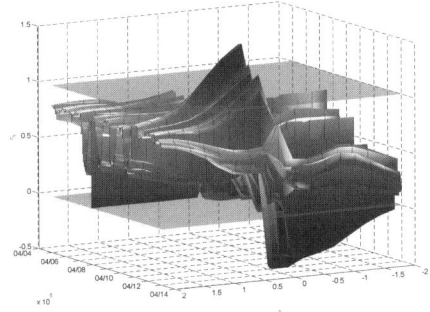

(d) 3D TES for incoming SrcPorts before and during refl. DDoS attack for $q = -2...2$. Diagonal axis: date (10 days), vertical axis: normalized entropy. Transparent layers: MIN and MAX at normal week days

Fig. 2. Reflector DDoS and DDoS 1

By a *spectrum pattern* we denote the consecutive c_q's for a representative set of values of q. In particular, we compute the pattern for $q = [-2, -0.5, 0, 0.5, 2]$. For instance, the pattern $--0++$ means that S_q is below threshold for $q = [-2, -0.5]$, above threshold for $q = [0.5, 2]$ and in the normal range for $q = 0$. The following table shows the spectrum patterns for the described attacks:

		Src IP					Dst IP					Src Port					Dst Port					AS			
q =		-2	-½	0	½	+2	-2	-½	0	½	+2	-2	-½	0	½	+2	-2	-½	0	½	+2	-2	-½	0	½	+2
Refl. DDoS	in	+	+	0	-	-	+	0	0	0	+	-	-	0	+	0	+	+	0	-	-	+	+	0	-	-
	out	+	0	0	+	0	+	+	0	-	-	+	+	0	-	-	-	-	0	+	0	+	+	0	-	-
DDoS 1	in	+	+	+	+	0	+	+	0	-	-	+	+	0	-	-	0	0	0	-	-	+	+	+	+	0
	out	0	0	0	-	-	+	+	+	+	0	+	+	0	-	-	-	-	0	-	-	+	+	+	+	0
DDoS 2	in	+	+	+	+	0	+	+	0	-	-	+	+	0	-	-	-	-	0	-	-	+	+	0	-	0
	out	0	0	0	+	0	0	0	0	0	0	+	0	0	0	0	0	0	0	-	-	0	0	-	-	-
Blaster W.	in	+	+	+	-	0	+	+	+	+	0	+	+	0	-	0	+	+	-	-	-	+	+	+	-	0
	out	+	+	0	0	0	+	+	+	+	0	+	+	0	-	0	+	+	0	-	-	+	+	+	-	-
Witty W.	in	0	0	0	-	-	+	+	+	+	0	+	+	0	-	-	+	+	+	+	+	0	0	0	+	+
	out	0	0	0	0	0	0	0	0	0	-	0	0	0	0	0	0	0	0	0	0	0	0	0	0	0

For each attack, incoming and outgoing traffic is considered separately. Selected features are src/dst addresses and ports as well as the AS numbers.[2]

The web servers used as reflectors in the refl. DDoS attack appear in the incoming DstIPs (requests from the real attackers). The number of reflectors (30,000) was large enough to increase the area of the high activity domain, resulting in a positive alert for $q = 2$. The relative activity of rare events was further reduced, amplifying their impact in the low activity domain and resulting in another positive alert for $q = -2$. The victim, being a single high activity host, had a contrary influence on the outgoing DstIPs and AS. The relative activity of other hosts was reduced by the appearance of the new heavy hitter and thus the overall area of the high activity domain was decreased. The reduction in relative activity also occurred to the already rare hosts, again amplifying their impact in the low activity domain. A similar effect is observed in the incoming DstPorts, where a concentration on port 80 is induced by the attack. However, the incoming SrcPorts where randomly distributed and activated virtually all ports. As a consequence, the former rare ports experienced a lift in activity and did not contribute to the low activity domain anymore, leading to negative alerts for $q < 0$. Figure 2(d) nicely illustrates the observed pattern $(--0+0)$. Note that the patterns are symmetric with respect to the diagonal. That is, changes in incoming SrcIP/SrcPort columns are reflected in outgoing DstIP/DstPort columns and vice versa. This indicates that the reflectors actually managed to reply to all requests (no egress filter was in place).

The main difference between the refl. DDoS and the ordinary DDoS attacks is that the former uses real hosts (the reflectors), whereas the latter uses massively spoofed source IP addresses. For both attacks, the incoming SrcIP TES was affected over a wide range $(++++0)$, including the SrcIP count $(q = 0)$. For the DDoS 2, however, the alerts in outgoing DstIPs is missing because no response flows were generated.

For both, the Blaster and the Witty worm, destination addresses of spreading attack traffic were generated randomly, much the same way as sources were spoofed during the DDoS attacks. And in fact, the pattern exhibited by incoming worm DstIPs is exactly the same as the pattern for incoming DDoS SrcIPs. The pattern produced by random feature selection $(++++0)$ is also visible in incoming DstPort for the Witty worm. On the other hand, the pattern specific to feature concentration $(++0--)$ is for instance visible in incoming Witty SrcPort (fixed to UDP 4000), incoming refl. DDoS DstPort (fixed to TCP 80) or incoming DstIPs for DDoS 1 and 2.

Random feature selection can have a different impact on ports than on IP addresses. Whereas incoming DstPort for Witty shows the typical pattern, the one for incoming SrcPorts of the refl. DDoS looks quite different $(--0+0)$. Random selection of IP addresses leads to many addresses with very low activity because the range of potential addresses is big. For ports, the range is limited to 65535 values. Thus, if intensive random port scanning is performed, all ports are often revisited and become frequent, basically eliminating the low activity area. This is what happened in the refl. DDoS case, indeed. We conclude that for ports, the strength (volume) of the attack plays a crucial role. For low volume attacks, the random port pattern looks like the random IP pattern, however, increasing attack volume shifts the pattern toward $--0+0$.

---

[2] Note that our traffic is recorded at a single stub AS. Consequently, source AS are shown for incoming and destination AS for outgoing traffic, respectively.

Summing up, we see that fundamental distribution changes such as concentration or dispersion of features are well reflected by different TES patterns and can therefore be used to infer underlying traffic structure. In future work, we will consider the effect of attack volume as well as additional patterns, e.g., the distribution of flow sizes and durations. The final goal is to develop a comprehensive and diverse set of TES patterns, suitable to accurately detect and classify network anomalies. For this, we need to do a more in-depth evaluation to prove that the improved detection sensitivity does not come along with a high ratio of false positives. Because our preliminary results suggest that TES is very robust (e.g., 8 days without a false alarm in 2(c)) even when using our trivial detection approach, we are positive that this will not be the case.

5   Related Work

Shannon entropy analysis has been applied successfully to the detection of fast Internet worms [6] and anomaly detection in general [7,8]. A different application of entropy is presented in [19], where the authors introduce an approach to detect anomalies based on Maximum Entropy estimation and relative entropy. The distribution of benign traffic is estimated with respect to a set of packet classes and is used as the baseline for detecting anomalies. In [9], Ziviani et al. propose to use Tsallis entropy for the detection of network anomalies. By injecting DoS attacks into several traffic traces they search for the optimal q-value for detecting the injected attacks. However, our results suggest that looking at a single time series for a specific value of q is not enough for revealing different types of anomalies. Furthermore, they do not look at negative values of q for which the entropy is very sensitive to changes in the low-activity region of the distribution. This might be linked to the fact that their evaluation is based on sampled or even anonymized traces. Truncation of 11 bits in IP addresses (as applied to the Abilene traces) might remove the formerly rare elements by aggregating them on the subnet level. However, aggregation is not necessarily a bad thing. Our results show that if multiple levels of aggregation such as IP addresses (fine grained) or Autonomous Systems (coarse grained) are used, aggregation turns out to be a powerful tool to reveal and classify anomalies.

6   Conclusion

The characterization and visualization of changes in feature distributions involves the analysis and storage of millions of data points. To overcome this constraint, we propose a new method called Traffic Entropy Spectrum. Our evaluation shows that the TES is very sensitive to changes that are small compared to the overall size of the observed network. Furthermore, we demonstrate that we can capture changes introduced by different types of anomalies using just a few Tsallis entropy values and find that our method does not require adaptation of its parameters even though the network and the underlying traffic feature distributions change significantly. On the detection side, we propose to use the information from the TES to derive patterns for different types of anomalies. We present ideas how we could use them to automatically detect and classify anomalies. In a next step, we plan to do a detailed analysis of the patterns of

different anomalies and cross-validate them with traces from various networks. This
will eventually enable us to develop a TES-based anomaly detection and classification
engine.

References

1. Barford, P., Kline, J., Plonka, D., Ron, A.: A signal analysis of network traffic anomalies. In:
   IMW 2002: Proceedings of the 2nd ACM SIGCOMM Workshop on Internet measurment,
   pp. 71–82. ACM, New York (2002)
2. Scherrer, A., Larrieu, N., Owezarski, P., Borgnat, P., Abry, P.: Non-gaussian and long memory statistical characterizations for internet traffic with anomalies. IEEE Transactions on De-
   pendable and Secure Computing 4(1), 56–70 (2007)
3. Dubendorfer, T., Plattner, B.: Host behaviour based early detection of worm outbreaks in
   internet backbones. In: 14th IEEE WET ICE, pp. 166–171 (2005)
4. Cisco Systems Inc.: Netflow services solutions guide, http://www.cisco.com
5. Quittek, J., Zseby, T., Claise, B., Zander, S.: Rfc 3917: Requirements for ip flow information
   export (ipfix) (October 2004)
6. Wagner, A., Plattner, B.: Entropy based worm and anomaly detection in fast ip networks. In:
   14th IEEE WET ICE, Linköping, Sweden (June 2005)
7. Lakhina, A., Crovella, M., Diot, C.: Diagnosing network-wide traffic anomalies. In: ACM
   SIGCOMM, Portland (August 2004)
8. Li, X., Bian, F., Crovella, M., Diot, C., Govindan, R., Iannaccone, G., Lakhina, A.: Detection
   and identification of network anomalies using sketch subspaces. In: Internet Measurement
   Conference (IMC), Rio de Janeriro, Brazil, pp. 147–152. ACM, New York (2006)
9. Ziviani, A., Monsores, M.L., Rodrigues, P.S.S., Gomes, A.T.A.: Network anomaly detection
   using nonextensive entropy. IEEE Communications Letters 11(12) (2007)
10. Shannon, C.: Prediction and entropy of printed english. Bell System Tech. Jour. (January
    1951)
11. Tsallis, C.: Possible generalization of boltzmann-gibbs statistics. J. Stat. Phys. 52 (1988)
12. Tsallis, C.: Nonextensive statistics: theoretical, experimental and computational evidences
    and connections. Brazilian Journal of Physics (January 1999)
13. Tsallis, C.: Entropic nonextensivity: a possible measure of complexity. Chaos (January 2002)
14. Dauxois, T.: Non-gaussian distributions under scrutiny. J. Stat. Mech. (January 2007)
15. Wilk, G., Wlodarczyk, Z.: Example of a possible interpretation of tsallis entropy. arXiv cond-
    mat.stat-mech (November 2007)
16. Willinger, W., Paxson, V., Taqqu, M.S.: Self-similarity and heavy tails: Structural modeling
    of network traffic. In: Statistical Techniques and Applications (1998)
17. Kohler, E., Li, J., Paxson, V., Shenker, S.: Observed structure of addresses in ip traffic. In:
    Proceedings of the SIGCOMM Internet Measurement Workshop, pp. 253–266. ACM, New
    York (2002)
18. SWITCH: The swiss education and research network, http://www.switch.ch
19. Gu, Y., McCallum, A., Towsley, D.: Detecting anomalies in network traffic using maximum
    entropy estimation. In: IMC 2005, pp. 1–6. ACM, New York (2005)

Author Index

Printing: Mercedes-Druck, Berlin
Binding: Stein+Lehmann, Berlin